The National Energy Modeling System

Committee on the National Energy Modeling System
Energy Engineering Board
Commission on Engineering and Technical Systems

in cooperation with the

Committee on National Statistics
Commission on Behavioral and Social Sciences and Education

NATIONAL RESEARCH COUNCIL

NATIONAL ACADEMY PRESS

Washington, D.C. 1992

NOTICE: The project that is the subject of this report was approved by the Governing Board of the National Research Council, whose members are drawn from the councils of the National Academy of Sciences, the National Academy of Engineering, and the Institute of Medicine. The members of the committee responsible for the report were chosen for their competencies and with regard for appropriate balance.

This report has been reviewed by a group other than the authors according to procedures approved by a Report Review Committee consisting of members of the National Academy of Sciences, the National Academy of Engineering, and the Institute of Medicine.

The National Academy of Sciences is a private, nonprofit, self-perpetuating society of distinguished scholars engaged in scientific and engineering research, dedicated to the furtherance of science and technology and to their use for the general welfare. Upon the authority of the charter granted to it by the Congress in 1863, the Academy has a mandate that requires it to advise the federal government on scientific and technical matters. Dr. Frank Press is president of the National Academy of Sciences.

The National Academy of Engineering was established in 1964, under the charter of the National Academy of Sciences, as a parallel organization of outstanding engineers. It is autonomous in its administration and in the selection of its members, sharing with the National Academy of Sciences responsibility for advising the federal government. The National Academy of Engineering also sponsors engineering programs aimed at meeting national needs, encourages education and research, and recognizes the superior achievements of engineers. Dr. Robert M. White is president of the National Academy of Engineering.

The Institute of Medicine was established in 1970 by the National Academy of Sciences to secure the services of eminent members of appropriate professions in the examination of policy matters pertaining to the health of the public. The Institute acts under the responsibility given to the National Academy of Sciences by its congressional charter to be an adviser to the federal government and, upon its own initiative, to identify issues of medical care, research, and education. Dr. Kenneth I. Shine is president of the Institute of Medicine.

The National Research Council was organized by the National Academy of Sciences in 1916 to associate the broad community of science and technology with the Academy's purposes of furthering knowledge and of advising the federal government. Functioning in accordance with general policies determined by the Academy, the Council has become the principal operating agency of both the National Academy of Sciences and the National Academy of Engineering in providing services to the government, the public, and the scientific and engineering communities. Dr. Frank Press and Dr. Robert White are chairman and vice-chairman, respectively, of the National Research Council.

This report and the study on which it is based were supported by Contract No. DE-ACO1-90PE79065 from the U. S. Department of Energy.

Library of Congress Catalog Card No. 91-68549
International Standard Book Number 0-309-04634-3
S-476

Additional copies of this report are available from:

National Academy Press
2101 Constitution Ave., N.W.
Washington, D.C. 20418

Printed in the United States of America

COMMITTEE ON THE NATIONAL ENERGY MODELING SYSTEM

PETER T. JOHNSON, *Chairman*, Former Administrator, Bonneville Power Administration, McCall, Idaho
DENNIS J. AIGNER, Dean, Graduate School of Management, University of California, Irvine
DOUGLAS R. BOHI, Director, Energy and Natural Resources Division, Resources for the Future, Washington, D.C.
JAMES H. CALDWELL, Jr., Consultant, Maryland
ESTELLE B. DAGUM, Director, Time Series Research Division, Statistics Canada, Ottawa, Ontario, Canada
DANIEL A. DREYFUS, Vice President, Strategic Planning and Analysis, Gas Research Institute, Washington, D.C.
EDWARD L. FLOM, Manager, Industry Analysis and Forecasts, Amoco Corporation, Chicago, Illinois
DAVID B. GOLDSTEIN, Senior Staff Scientist, Natural Resources Defense Council, San Francisco, California
LOUIS GORDON, Professor, Department of Mathematics, University of Southern California, Los Angeles
VELLO A. KUUSKRAA, President, Advanced Resources, Arlington, Virginia
JAMES W. LITCHFIELD, Director of Power Planning, Northwest Power Planning Council, Portland, Oregon
STEPHEN C. PECK, Director, Environment Division, Electric Power Research Institute, Palo Alto, California
MARC H. ROSS, Professor, Department of Physics, University of Michigan, Ann Arbor
EDWARD S. RUBIN, Professor, Departments of Engineering and Public Policy and Mechanical Engineering, and Director, Center for Energy and Environmental Studies, Carnegie-Mellon University, Pittsburgh, Pennsylvania
JAMES L. SWEENEY, Chairman, Department of Engineering-Economic Systems, and Director, Energy, Natural Resources, and the Environment Program, Center for Economic Policy Research, Terman Engineering Center, Stanford University, Stanford, California
DAVID O. WOOD, Director, Center for Energy Policy Research, Sloan School of Management, Massachusetts Institute of Technology, Cambridge

Staff

MAHADEVAN (DEV) MANI, Director, Energy Engineering Board and Study Director (until February 1991)
JAMES ZUCCHETTO, Senior Program Officer and Study Director (as of February 1991)
JUDITH AMRI, Administrative Associate
PHILOMINA MAMMEN, Senior Project Assistant
ANN COVALT, Consulting Editor

ENERGY ENGINEERING BOARD

JOHN A. TILLINGHAST, *Chairman*, Tiltec, Portsmouth, New Hampshire
DONALD B. ANTHONY, Consultant, Houston, Texas
RICHARD E. BALZHISER, Electric Power Research Institute, Palo Alto, California
BARBARA R. BARKOVICH, Barkovich and Yap, Consultants, San Rafael, California
JOHN A. CASAZZA, CSA Energy Consultants, Arlington, Virginia
RALPH C. CAVANAGH, Natural Resources Defense Council,
  San Francisco, California
DAVID E. COLE, University of Michigan, Ann Arbor, Michigan
H. M. HUBBARD, University of Hawaii, Honolulu, Hawaii
ARTHUR E. HUMPHREY, Lehigh University, Bethlehem, Pennsylvania
  (to February 1991)
CHARLES IMBRECHT, California Energy Commission, Sacramento, California
CHARLES D. KOLSTAD, University of Illinois, Urbana, Illinois
HENRY R. LINDEN, Gas Research Institute, Chicago, Illinois
JAMES J. MARKOWSKY, American Electric Power Service Corporation, Columbus, Ohio
  (to February 1991)
SEYMOUR L. MEISEL, Mobile R&D Corporation (retired), Princeton, New Jersey
DAVID L. MORRISON, The MITRE Corporation, McLean, Virginia
MARC H. ROSS, University of Michigan, Ann Arbor, Michigan
MAXINE L. SAVITZ, Garrett Ceramic Component Division, Torrance, California
HAROLD H. SCHOBERT, The Pennsylvania State University, University Park,
  Pennsylvania
GLEN A. SCHURMAN, Chevron Corporation (retired), San Francisco, California
JON M. VEIGEL, Oak Ridge Associated Universities, Oak Ridge, Tennessee
BERTRAM WOLFE, General Electric Nuclear Energy, San Jose, California

Staff

ARCHIE L. WOOD, Executive Director, Commission on Engineering and Technical
  Systems, and Director, Energy Engineering Board (to January 1991)
MAHADEVAN (DEV) MANI, Director, Energy Engineering Board (as of January 1991)
KAMAL ARAJ, Senior Program Officer
GEORGE LALOS, Senior Program Officer
JAMES ZUCCHETTO, Senior Program Officer
JUDITH AMRI, Administrative Associate
THERESA FISHER, Senior Project Assistant
PHILOMINA MAMMEN, Senior Project Assistant

DEDICATION

This report is dedicated to the memory of David O. Wood, a good friend and colleague. David contributed greatly to energy and economic modeling through his commitment to conceptual rigor and open debate, and especially through his own intellectual integrity. He served as a member of the Committee on the National Energy Modeling System until April 1991.

CONTENTS

EXECUTIVE SUMMARY   1
   Principal Findings, 2
      Principal Recommendations, 3
   Timing of NEMS Development, 3
   NEMS Management, 3
   NEMS Design, 4

1 INTRODUCTION   9
   Study Genesis and Background, 10
   Scope of the Study, 10
   Organization of the Report, 13

2 NEMS REQUIREMENTS   15
   The Role of Models in Policy Analysis and Planning, 15
     Energy Policy-Related Models, 15
     General Approaches to Modeling, 17
     The Benefits of Models, 18
     The Limitations of Models, 19
   The Mission and Functions of DOE and EIA, 20
     Strategic Analysis, 20
     Data Collection and Information Dissemination, 21
     R&D Program Planning, 21
     Current Modeling Capabilities Within DOE and EIA, 22
   NEMS in the Broad Context of Policy Analysis, 23
     Keeping NEMS Outward Looking, 24
     Capabilities Outside DOE, 26
   The National Energy Strategy Experience, 27
     Overview of the NES Exercise, 27
     Committee View of the NES Analysis, 28
   Directions in Energy Modeling, 31
     General Trends, 32
     Quantification of Uncertainty, 34
     Long-Term Forecasting, 36
   NEMS Requirements, 38
   Findings and Recommendations, 42

3 NEMS ARCHITECTURE   45
    Overview, 45
    Modular Architecture, 48
       Advantages of a Modular System, 49
       Disadvantages of Modular Systems, 50
    Integrating Model Operation, 50
    Proposed Modules, 55
       The Control Module, 56
       Fossil Fuel Supply Modules, 56
       Energy Conversion Modules, 58
       Renewable Energy Conversion Modules, 59
       The International Energy Module, 61
       The Interindustry Growth Model, 63
       Energy Demand Modules, 64
       Reduced-Form Models, 74
       Satellite Modules, 75
       Report Writers, 77
    NEMS Compared to Current DOE Modeling, 78
    NEMS Treatment of Conceptual Issues, 79
       Market Disequilibrium, 79
       Uncertainty, 80
       Contingent Strategies, 81
       The Formation of Expectations, 82
       Environmental Constraints, 84
    Operational Issues for NEMS Development, 85
    Recommendations, 87

4 IMPLEMENTATION OF NEMS   89
    Lead Organization for the NEMS, 89
    Suggestions for Implementation, 90
    Employee Environment, 91
    Management of NEMS Development, 93
    Motivation and Usefulness, 95
    Recommendations, 95

APPENDIXES

A-1	Scope of Work	97
A-2	Committee Charge	101
B	First Advisory Report	103
C	The Mission and Functions of the Department of Energy	113
D	Illustrative Case Studies	119
E	A Brief Description of DOE and EIA Models	129
F	Meetings and Activities	135

REFERENCES AND BIBLIOGRAPHY   141

LIST OF TABLES

2-1	DOE Policy Issues	30
2-2	Environmental Issues Relevant to NEMS	39
3-1	Available Data and Information, Industrial Energy Demand	71

LIST OF FIGURES

2-1	Scheme of the National Energy Analysis System and the EIA's scope within it	24
2-2	Scheme of the interface between the NEMS and the National Energy Analysis System	25
3-1	Simple representation of proposed NEMS architecture	47
3-2	Illustration of the convergence of price to a supply-demand equilibrium	52
3-3	Illustration of multiple equilibria	53
3-4	Proposed NEMS architecture in greater detail	55
3-5	Modeling a system-wide environmental constraint	86

ACKNOWLEDGMENTS

The committee gratefully acknowledges the help of the following individuals: Secretary of Energy James D. Watkins; Deputy Secretary of Energy W. Henson Moore; Energy Information Administrator Calvin Kent; Abraham Haspel, Robert C. Marlay, Eric Petersen, Peter Saba, and Linda Stuntz, U.S. Department of Energy; Linda Carlson, John Conti, Ronald Earley, Bob Eynon, Ed Flynn, Douglas R. Hale, John Holte, Mary J. Hutzler, W. Calvin Kilgore, Erik Kreil, Fred Mayes, John Pearson, Lawrence A. Pettis, Susan Shaw, C. William Skinner, and Scott Sitzer, Energy Information Administration; Sharon Belanger and Roger Nail, AES Corporation; Peter Blair, John Gibbons, and Henry Kelly, Office of Technology Assessment; Jae Edmonds, Batelle Pacific Northwest Laboratories; David Gray, David Morrison and Glen Tomlinson, Mitre Corporation; Susan Hickey, Bonneville Power Administration; Eric Hirst, Oak Ridge National Laboratory; Philip Hummell, Rich Richels, and Colleen Hyams, Electric Power Research Institute; Mark Inglis, ICF Resources; Ralph L. Keeney, University of Southern California; Dan Kirshner, Environmental Defense Fund; Lester Lave, Carnegie-Mellon University; Terry Morlan, Northwest Power Planning Council; Dale Nesbitt, Decision Focus, Inc.; Daniel Nix, California Energy Commission; Andrew Plattinga, Resources for the Future; and Miron Straf and Eugenia Grohman, Commission on Behavioral and Social Sciences and Education, National Research Council.

EXECUTIVE SUMMARY

In response to a request from the U.S. Department of Energy (DOE), in the summer of 1990 the National Research Council (NRC) established the Committee on the National Energy Modeling System (NEMS) to advise DOE and the Energy Information Administration (EIA) on the development and application of a modeling system to support energy policy analysis and strategic planning. As discussed later, this system should have the capability to simulate the effects of various energy policies on U.S. energy supply and demand including economic, environmental and national security impacts. In this report, the committee offers its findings and recommendations on the process of developing a NEMS, the architecture and data needs of the system, and the organizational and management actions that will, in the committee's opinion, help make the system work. The executive summary identifies the most important of these findings and recommendations.

The committee was briefed by DOE and EIA on their analytical and modeling activities in policy analysis for the National Energy Strategy (NES). The committee was also briefed on DOE and EIA modeling capabilities that might play a role in future NEMS development. In addition, the committee reviewed state-of-the-art energy modeling and analysis through current publications and invited presentations by individuals and public and private organizations engaged in such work.

The report and its findings and recommendations rest on this information and the committee's deliberations. Principal findings of the committee are as follows:

PRINCIPAL FINDINGS

o **The set of EIA models reviewed by the committee at the beginning of this study constitutes a reasonable starting point for developing a National Energy Modeling System. However, considerable development will be needed to attain a modeling system satisfying the requirements outlined in this report.**

The committee believes that modeling can play a valuable part in energy policy analysis. In its first advisory report issued January 30, 1991, the committee observed that DOE has no comprehensive model or set of models that can respond adequately to the needs of the NES but "the approach taken by DOE in using available models, and off-line supplemental analysis as necessary, was a rational response to the department's need for expedient support of the NES process" (NRC, 1991a; reprinted as Appendix B to this report; also see Appendix E). The committee noted, however, that the set of models used by DOE in the NES had "significant limitations relative to the analytical results reported by DOE," and that one needs "to appreciate the limited power of the existing set of models used for evaluating policy choices."

Evaluation of the NES effort thus indicated that the NEMS must significantly exceed the capabilities of existing DOE models. The committee recommends a course of action that builds on existing capabilities.

o **The NEMS Program, once established, should complement and interact with a variety of other public and private groups that contribute to policy analysis.**

The development and use of models is only one aspect of policy analysis. Policy analysis also includes the identification of policy issues and initiatives of importance, development of assumptions, validation of models and data, interpretation of model results, and attention to social values. Additionally, the capabilities of the NEMS will complement many credible public and private models and analysts.

o **Successful development of the NEMS will require the Secretary of Energy and EIA Administrator to establish and foster an organizational environment that is outward-looking and ensures greater intellectual and institutional commitment to its development and maintenance.**

After more than a decade of budget stringency and less than full support by the DOE at large, DOE/EIA "culture" has clearly not moved forward. For the NEMS to fulfill its potential as a national energy modeling and data resource system, DOE and EIA must create a supporting institutional environment that is sufficiently rich, broad, and interactive, that relates well to the DOE's policy and program offices and to appropriate federal, state, regional, and nongovernment groups.

The committee's principal recommendations follow here; additional recommendations are offered throughout the report. The committee's goal is to lay the foundations for creating a NEMS and associated institutional environment to support energy policy analysis that is objective, useful, and of the highest professional standards. The recommendations are classified by three major issues for the NEMS: the timing of its development, its management, and its design.

PRINCIPAL RECOMMENDATIONS

Timing of NEMS Development

1. **The EIA and DOE should move quickly to configure an initial National Energy Modeling System within the next one to two years and apply it to policy issues, including the next National Energy Strategy.**

The most useful and cost-effective approach to begin the development of NEMS is to build on existing models, modified as appropriate, to address the most important national energy policy problems. Using this approach, NEMS can be implemented quickly and provide analysis for decisions in the next NES. NEMS does not need to be fully developed to be a valuable near-term tool. The recommended structure of NEMS will facilitate later system improvements by allowing the substitution or addition of new or modified modules. Through constant use and this evolutionary approach, NEMS will become increasingly more comprehensive, accurate, and useful.

The committee is concerned that if the pace of NEMS development doesn't allow its early use, then an important opportunity will be missed to improve the analysis of energy policy issues important to DOE. If priority is given to NEMS in the allocation of budgets and personnel, an initial system can be configured and applied in the next NES exercise. By virtue of its application, the system will receive greater peer review and undoubtedly will be changed and improved and expanded for future use.

NEMS Management

2. **The Secretary of Energy should designate the EIA Administrator as the chief executive for the implementation of the NEMS, and make this assignment one of the main performance requirements for the Administrator.**

By law, EIA was established within DOE as an independent, nonadvocacy group to collect, analyze, and report data for the federal government. (In contrast, DOE's Office of Policy, Planning and Analysis advances the philosophy of the Department.) Thus, EIA is insulated from the vagaries of the policy process and political pressures to which the rest of the DOE is exposed. EIA is traditionally looked to for impartial analyses of energy supply and demand, for example, as reported in their *Annual Energy Outlook*. Such independence helps confer greater objectivity and credibility to EIA activities.

For the successful development of NEMS, the committee believes that the Secretary needs to vest the responsibility for its development and implementation directly with the EIA

Administrator. The Administrator should be directly responsible because the development of NEMS will affect most other agency activities, will require new functions, and will overlap with functions of other DOE offices.

> 3. **Toward keeping NEMS outward-looking and sensitive to relevant policy issues, EIA should form a Users Advisory Council of likely NEMS users from within DOE, other federal, state, and regional agencies, and private organizations. EIA should develop and manage the NEMS with the intensive involvement of this advisory group.**

User advisory groups are vital to the effective functioning of NEMS. An important use of NEMS will be for energy policy analysis by those outside of EIA. To help ensure the responsiveness of NEMS to users' needs, EIA must maintain close working relationships with potential users, both inside and outside DOE, with experience in policy analysis. Various federal and nonfederal organizations have significant data resources and modeling capabilities. These groups, resources, and advice should be sought and used well in NEMS development.

> 4. **DOE should take action to attract and retain highly skilled professionals for the design, development and implementation of the NEMS.**

The initiation and sustained development of NEMS will require talent. The models developed will continually require updating and maintenance. NEMS will be valuable only if supported on a continuing basis with adequate budgets and skilled personnel. The human resources now devoted to NEMS are seriously limited, especially the number of senior personnel.

NEMS Design

> 5. **NEMS should be designed to estimate and display as primary outputs the economic, environmental, and national security implications of alternative energy policies, with priority given first to the mid-term time horizon and then to the longer term.**

NEMS should produce economic, environmental, and national security measures relevant to the analysis of public policy. Decision makers need more information than energy quantities and prices over time. Economic outcomes include consumers' and producers' surpluses, gross national product (GNP), and the federal budget deficit. Environmental measures include the direct and indirect emissions of major pollutants and their effects. Security measures include the economic impacts of energy supply interruptions. Regional disaggregation and equity measures are also relevant to some policy analyses. Clear accessible graphical outputs are also essential to users.

Three roughly defined time horizons are appropriate for the analysis of energy policy issues: the short-term (roughly up to two years), mid-term (up to about 25 years), and long-term (beyond 25 years). The committee believes that NEMS should initially focus on developing mid-term analytic capability: this time horizon is of greatest relevance for most important national policy issues. Priority should then be placed on the development of long-term models, which do not now currently exist at EIA.

6. NEMS must incorporate the behavioral and policy-driven aspects of decisions about energy use, fuel and technology choices, and energy supply investments.

Many of the important energy policy issues for the foreseeable future will concern energy consumption and policy options to influence energy consumption. To evaluate the potential effectiveness of economic incentives and disincentives and regulations, the nature of relevant decision-making processes must be reasonably reflected in the models. To a very great extent, decisions about energy consumption and even about investments in energy supply are now being described in models based only upon the surmise of the modeler. Resources should be devoted to capturing available descriptive information on the behavior of energy users and suppliers in model structures and to supporting related research and collection of further information.

7. NEMS should be designed to represent and analyze the effects of uncertainty explicitly.

Inherent with energy modeling are many types of uncertainties, which are often not sufficiently addressed in policy analysis. By characterizing such uncertainties and their consequences, NEMS can help to identify and manage the risks associated with energy policy alternatives through the design of insurance or hedging strategies.

Two major types of uncertainties that should be considered initially in NEMS are (1) uncertainties in model input parameter values, which determine projections of future energy use (e.g., parameters describing economic growth, energy prices, and technological change); and (2) uncertainties in the basic relationships conjectured between energy use and economic or behavioral activity (e.g., the structural or causal relationships assumed by NEMS components). To address the first type of uncertainty, sensitivity analysis and scenario studies should be performed to assess the implications of alternative assumed situations on key model output measures. To address the second type of uncertainty, NEMS's modular structure can be exploited to analyze the implications of alternative model formulations for various NEMS components (including those developed outside DOE and EIA).

Over the longer term, other types of uncertainty analysis should also be incorporated in NEMS. For example, the development of reduced-form modules (simplified representations of more complex model components) will facilitate the use of probabilistic methods that allow uncertainties to be expressed as probability density functions describing the likelihood of particular outcomes (in response to assumed uncertainties in model inputs). Such methods already are in use (e.g., by the Northwest Power Planning Council) and have provided valuable insight. They have also been used in research and development planning.

8. NEMS should be designed to be modular in structure. The system should readily accommodate substitution of alternative models (modules).

The committee proposes a modeling system that would consist of a set of modules linked together by a simple control module. The modules should be designed so that they can be run separately, all together, or in combinations, depending on the analytical need. Each component module would provide the richness of detail and diversity of structure appropriate to the sector it represents. The control module would provide a simple set of switches to govern the operation of the component modules and an algorithm to achieve equilibrium of supply, demand, and price. The structure proposed represents a logical continuation of modeling development within EIA.

Modular architecture would facilitate decentralized development and maintenance of the modules. Once the inputs, outputs, and interfaces between models are appropriately defined and specified, then development of the modules could proceed in a decentralized manner.

Modular architecture would allow the substitution of alternative modules embodying different conceptual structures, theories, empirical representations, and data bases to examine some of the sources of forecast uncertainty. This would allow the use of existing models from groups outside the DOE and EIA, especially with regard to demand-side modeling.

As mentioned in the discussion above on uncertainty, reduced-form versions of the component modules should be developed that will simplify the more complete modules by approximating the relationship between input variables and output variables. Reduced-form modules would increase the transparency of the full modules by allowing users to readily perceive their aggregate behavior.

The committee believes that at least the reduced-form version of NEMS should be configured to run on microcomputers or personal computers.[1] If the system could run on widely available hardware configurations, such as personal computers, the number of analysts who could use the models would be greatly expanded. Such access would increase the numbers who understand, use, and test the system. This hardware restriction is essential to achieving the goals of transparency and quick turnaround. Alternatively, but less desirable, NEMS could be configured to run on other widely available hardware configurations (e.g. workstations) if it were coded using a machine-independent computer language.

---

[1] The committee uses the terms "personal computer" and "microcomputer" as generic terms for small computers that are widely available to a variety of users. They could include computers made by IBM or others using the DOS software, Apple, NEXT computers, or any other widely available computers.

9. NEMS architecture should provide for quick turnaround. EIA procedures must support this rapid response capability.

Although the EIA will develop, manage, and maintain NEMS, and use it for its own purposes whenever appropriate, the DOE Office of Policy, Planning and Analysis and program offices will need to access the system independently. These offices must run some analyses very quickly (in hours rather than days, including hard copy output). Recognizing tradeoffs between modeling level of detail and run time, analysts can use the reduced-form models mentioned earlier extensively in this process. The committee suggests that for such policy analysis, computer outputs could be generated without an EIA imprimatur; only the results of final runs might be reviewed through the full EIA administrative process before their release in policy documents or reports.

10. EIA should create a group to develop a long-term modeling and analysis capability.

The mid-term NEMS modeling effort that the committee recommends be implemented immediately should not simply be extrapolated for long-term modeling (for the time horizon beyond 25 years). Instead, the EIA should create a distinct group to develop long-term modeling and analysis capability. In addition to relating this long-term capability to the medium-term modeling system in a consistent fashion, this group should avail itself of expertise of external individuals and groups who already focus on long-term energy modeling. This effort should be initiated quickly. However, the actual development or acquisition of long-term modeling capability should be given lower priority among the tasks recommended for the NEMS effort.

11. A major effort is needed to collect more extensive data and information on the U.S. energy system, especially on end use.

Accurate and relevant data are needed to support the development of models for the NEMS. But current EIA data efforts lean too heavily toward mid-term supply-side data. Considerably more demand-side data collection is needed, as is the collection of long-term supply-side data. Use of DOE and EIA data bases should be supplemented by the use of data bases from other federal agencies, state and regional governments, and private organizations. Where appropriate, existing EIA surveys could also be modified to collect new kinds of data without major additional effort. Finally, new data-collection efforts on a one-time or continuing basis may be needed to meet reasonable modeling requirements.

A major effort is needed to collect disaggregated data on (1) the characteristics of both existing and new energy-using equipment; (2) the stock of energy-using equipment; (3) energy use by "service" category (e.g. space heating); (4) activities that underlie the various categories of energy use, such as the number and types of housing units in use; and (5) the cost and savings of technologies that can be implemented or developed to improve energy-efficiency (in modeling, "supply curves of conserved energy"). These data and the models that employ them need to be adequate to evaluate demand-side policies, such as schemes to encourage investment in efficient equipment or regulations mandating efficient performance of equipment. Such data and models are not adequate at present.

To support longer term energy modeling, new approaches and data are needed to (1) assess undiscovered resources in DOE resource assessments, (2) appropriately account for the large unconventional oil and gas resources into the assessment, (3) estimate the effects of continued evolution in extraction technology and production costs, and (4) evaluate the future contributions of coal, nuclear, and renewable energy, taking into account their economics and environmental implications.

Data requirements for renewables are different from those for fossil fuels. The difference between discovered reserves and undiscovered resources has no parallel for renewable resources. The modeling problem for renewable resources is lack of data on early adoption potential in small low-cost resource pockets and small high-value market niches.

The nation will be spending several trillion dollars over the next thirty years at satisfying our increasing demands for energy services. Federal investments in technologies for both supply and end use will be evaluated based not only on cost effectiveness but also on environmental and national security concerns. If a better modeling system can guide energy investments in a way that addresses environmental and national security concerns and reduces costs by only a percent, it will have paid for itself a thousand-fold. The remaining chapters of this report lay out the requirements, a proposed architecture, and a strategy for management that the committee feels will result in a modeling system that will respond to the nation's needs.

1

INTRODUCTION

The National Research Council (NRC) Committee on the National Energy Modeling System (NEMS) was established in the summer of 1990, to advise the U.S. Department of Energy (DOE) and the Energy Information Administration (EIA) on the long-term development of a modeling system to support national energy policy analysis and strategic planning.

In reviewing the original scope of work (Appendix A-1) from the DOE that initiated the study, the committee expected that there would at least be in use at DOE a preliminary version of a system of models definable as the National Energy Modeling System (NEMS). The committee also expected that there would be preliminary plans prepared by DOE for the further development of the NEMS, which could serve as a point of departure for the committee's work. At the first meeting of the committee on July 31, 1990, it became apparent that there was no NEMS per se or definitive plans for its development. The committee learned that in response to the demands of the National Energy Strategy (NES) exercise that was underway, DOE with the assistance of the EIA and others had configured an ensemble of models to support the NES analysis.

The committee requested and received presentations from DOE and EIA on this ensemble of models and their application to a representative slate of policy issues that were being considered in the NES exercise. The committee was also briefed on a preliminary set of requirements that EIA was developing for a NEMS, and subsequently EIA presented a comparison of those requirements to existing capabilities (involving models, data, and analysis) at EIA.

On the strength of those briefings, and reviews of documents and reports prepared by DOE, EIA and the national laboratories, the committee assessed, in broad terms, the efficacy of the models configured by DOE to support the NES activities. At the request of the Secretary of Energy, the committee's assessment was documented in its first advisory report issued in January 1991 (NRC, 1991a; Appendix B). In light of the foregoing circumstances and what the committee learned, the committee reinterpreted its charge to best utilize its resources (see Appendix A-2) and included this charge as part of its first advisory report. The committee also met with the Secretary of Energy and apprised him of its findings, including its reinterpreted charge. Subsequently, starting with its fourth meeting in January 1991, the committee focused its efforts exclusively on the development of a NEMS, the subject of this report.

This final report by the committee addresses NEMS applications and requirements, proposes a modeling architecture to satisfy these requirements, and suggests strategies for NEMS implementation.

STUDY GENESIS AND BACKGROUND

In a statement before the Senate Committee on Energy and Natural Resources, July 26, 1989, the Secretary of Energy, Admiral James D. Watkins, explained President Bush's plan for development of the NES. The Secretary directed the Deputy Under Secretary for DOE's Office of Policy, Planning and Analysis to coordinate the development of the first edition of the NES, issued in February 1991 (U.S. DOE, 1991a).

Toward developing the requisite data, analytical tools, and forecasting capabilities for the NES, the Secretary also announced that DOE would seek the advice of the National Academy of Sciences on how best to proceed with NEMS development.

> I have asked the National Academy of Sciences to examine our plans for the development of the NEMS and ensure that it will, to the maximum extent possible, address the critical energy issues before us. These include major environmental issues, strategic considerations and technology research and development...
>
> NEMS development will be an ongoing effort. It will probably take several years to improve DOE's modeling capability. I am determined that the NEMS will become the best modeling system that we can employ to analyze the issues facing us.

SCOPE OF THE STUDY

When this study began, one of its goals was to examine the ability of the existing DOE modeling system to analyze issues for energy policy and forecasts. The committee also broadly addressed the role of DOE's modeling in the development of the 1991 NES. These models include EIA supply and demand models and a few others, such as Fossil2, a systems dynamics model of the entire U.S. energy system, which was used by DOE's Office of Policy, Planning and Analysis to integrate the results and construct forecasts.

In assessing DOE models' support of NES activities, the committee considered the adequacy of their underlying data, assumptions, and methodology. In particular, the committee was concerned about the deterministic nature of the models used in the NES in making longer term projections, the limitations of Fossil2, the adequacy of demand-side data, and the lack of recognition and treatment of uncertainty in the NES analysis. The committee's comments about the uncertainty of projections encouraged the authors of the 1991 NES report to address this important issue explicitly, including on report graphs (U.S. DOE 1991a).

In the initial phase of its work that concluded with its first advisory report, the committee heard many hours of presentations from DOE and EIA and asked many questions about the methods and models used to develop the first National Energy Strategy. No definition of the NEMS was explicitly stated by the DOE. The committee concluded that the scope and purpose for the system could be derived from the statements of DOE managers and analysts and from the recent experience with the analytical activities that supported the NES effort that was being completed as the committee began its deliberations. The NEMS was clearly viewed by DOE's management to be a comprehensive, policy-oriented, modeling system with which the existing situation and alternative futures for the U.S. energy system could be described within the global energy context. The NEMS was expected to be applied in evaluating the potential impact of a broad range of policy initiatives upon the national energy outlook in measures that are relevant to national objectives, such as environmental management and energy security.

At that time, and based upon the above view of what constitutes a NEMS, the committee concluded that the ensemble of DOE and EIA models used in the NES exercise did not constitute an adequate National Energy Modeling System. Furthermore, there were no definitive plans laid out by DOE for further development of a NEMS. As mentioned above, given the mismatch between the orginial scope of work for the committee and the reality of the situation, the committee in its first advisory report reinterpreted its scope of work and defined a committee charter, which it presented to the Secretary of Energy (NRC, 1991a; Appendix B). The committee further concluded that if more detailed evaluations of existing models were desired, such work could be more appropriately conducted by outside groups (see Appendix A-2, the committee charge: "detailed technical assessments will not be undertaken of specific models currently in use at DOE or EIA").

The committee determined that its greatest contribution would be in first assessing the requirements of and then recommending the architecture for a future NEMS. Therefore, after the publication of its first advisory report the committee looked forward to the development of such a NEMS. The committee believes that the NEMS described in this final report embodies the characteristics needed to analyze critical national energy issues and to provide support for the ongoing NES efforts.

In a more specific sense, the suite of models that had been assembled to support the NES analysis that are housed in the EIA, elsewhere in the DOE, and in various national laboratories had come to be thought of as the prototype of the NEMS. This suite of models was, in fact, being evaluated within the DOE concerning its capabilities and shortcomings as if it were the formative NEMS.

The committee, therefore, referred to three sources of information in deriving an operating definition of the NEMS concept. The first is the ongoing analytical requirement to support the mission of the Department. While the NEMS is not, and should not strive to become, the sole modeling capability for the DOE's extensive analytical activities, the committee envisioned a role for the NEMS in directly supporting strategic policy formation and indirectly informing, coordinating, and disciplining analysis done with other decentralized modeling capabilities throughout the DOE.

The second source of information contributing to the committee's definition of the NEMS was the contemporary experience with the NES analytical support. The NES is a unique experiment in using analysis to address national energy issues in a comprehensive way. It provided a specific set of examples of the types of policy questions being posed for analysis and of the measures of value that are being assigned priority. The NES effort, therefore, presented the committee with a catalog of policy issues of importance and an empirical evaluation of the existing modeling capability to address them.

Finally the committee brought to the study the collective views of its members, who have substantial experience with public policy formulation and with applications of analytical and modeling techniques to that process. The current state of the art and the practice elsewhere throughout the analytical community were considered in comparison with the DOE approach and capability.

Based upon these three sources of information concerning what the NEMS could and should be, the committee concluded that the existing modeling capability within the DOE did not meet the requirements implied by management aspirations, particularly as revealed by the NES experience, and did not reflect what external evidence suggested could be provided within the practical constraints of time and resources.

This report develops a set of requirements for a NEMS as derived from the committee's consideration of the above sources of information. It prescribes the architecture of a modeling structure that can serve the purpose of the NEMS and that can evolve with the concept in the years ahead. It also discusses the approach to the implementation of a NEMS as set forth in the requirements and architecture with consideration for the work already in progress within the DOE and the practical limitations on resources that must be realized.

The appendices to the report also include substantial additional discussion of the background for the committee's deliberations.

ORGANIZATION OF THE REPORT

The following chapters address the policy analysis environment in which the NEMS would operate, propose an architecture for the system, and recommend actions for its successful implementation. In particular, Chapter 2 discusses the policy analysis environment and the needs of the DOE. It also briefly describes the capabilities and modeling activities within and outside DOE, the NES experience, and requirements that the NEMS should satisfy. Chapter 3 then proposes an appropriate architecture and components, or modules, for the NEMS, identifies corresponding data and information requirements, compares the proposed system to present DOE modeling, and discusses some conceptual issues for the proposed modeling approach. Finally, Chapter 4 addresses issues for the successful implementation of NEMS.

Beyond the committee's original scope of work in Appendix A-1 and charge in Appendix A-2, this report's appendixes include a reprint of the committee's first advisory report (Appendix B), additional discussion on DOE's mission and functions (Appendix C), brief committee case studies exploring the ways that NEMS could be employed (Appendix D), some further details about past and current EIA and DOE models (Appendix E), and a list of presentations to the committee (Appendix F).

2

NEMS REQUIREMENTS

This chapter describes the critera that a National Energy Modeling System (NEMS) should meet to be of greatest value in energy policy analysis, strategic planning and decision making. Based on these requirements, an architecture for the NEMS is recommended in Chapter 3.

The committee's view of NEMS requirements was shaped by five major issues that are addressed in this chapter: (1) the role of policy models and their strengths and weaknesses; (2) the mission and functions of the U.S. Department of Energy (DOE) and the Energy Information Administration (EIA); (3) the current modeling capabilities of DOE/EIA; (4) the National Energy Strategy (NES) experience as it bears on NEMS requirements; and (5) future general directions in energy modeling. These considerations provide the primary basis for the NEMS requirements summarized at the end of this chapter.

THE ROLE OF MODELS IN POLICY ANALYSIS AND PLANNING

Energy Policy-Related Models

Energy policy models can provide important information for national decision making (Ayres, 1978; AES, 1990; Hogan and Weyant, 1983; House and McLead, 1977; Koreisha and Stobaugh, 1979; Manne et al., 1985; Sweeney and Weyant, 1979). In fact, as stated by the committee in its first advisory report: "...the committee believes that models can play a crucial role in enabling informed judgments and decisions to be made in matters of national energy policy. Thus, the committee considers it vital that DOE continue to develop and sustain capabilities for analyzing national energy issues using resources from within the Department and from appropriate organizations in both the public and private sectors. President Bush articulated three primary areas of interest for energy policy in his introduction to the 1991 National Energy Strategy (NES): economic prosperity, environmental quality, and national security. The NES represents a new national effort, he said, to achieve "balance among our increasing need for energy at reasonable prices, our commitment to a safer, healthier environment, our determination to maintain an economy second to none, and our goal to reduce dependency by ourselves and our friends and allies on potentially unreliable energy supplies" (DOE, 1991a).

These aims are all widely accepted. Yet despite such general agreement, federal energy policy making has been marked by deep disagreements among different interests and little movement forward on significant issues. The disagreements have concerned the means of achieving desired ends, the precise nature of the goals, and the forms of acceptable tradeoffs among sometimes conflicting objectives.

High quality energy modeling can help bound these debates and move policy forward in two ways. First, modeling can provide insights to decision makers about the likely results of different policies. Second, it can help focus debates on scientific rather than ideological questions. Thus, energy modeling can help policy makers achieve greater consensus on proposed solutions and increase the likelihood of their adoption. Such model-assisted decision making has been used effectively in state and regional contexts over the last ten years.

One important kind of energy decision is whether to invest in new supply-side energy resources, such as oil wells, coal mines, and power plants, or in end-use efficiency measures, such as better insulation for refrigerators. Models can provide quantitative measures to help guide the understanding of the economic, environmental, and security implications of alternative energy investments. They also can help to estimate the results of different types of public policies (such as tax incentives to producers, transporters, and consumers, or payments to utilities for improving end-use efficiency). Some of the policy options that DOE and others considered in the 1991 NES activity are covered later in this chapter to further illustrate this point.

In considering the types of energy issues that models can help address, the committee distinguished three time horizons relevant to policy making: the short-term (up to about 2 years), the medium-term (up to about 25 years), and the long-term (beyond 25 years). For each time horizon a different modeling approach is needed.

Policy actions directed at short-term issues are likely to be more reactive than proactive. Short-term models therefore must address such problems as disequilibrium phenomena and the macroeconomic effects of energy disruptions (e.g., embargoes of energy supplies or sudden price hikes).

As explained below, most energy policy issues of national concern involve the medium term (up to about 25 years), such as the evaluation of alternative energy investments whose effects will not materialize in the short term. The emphasis in these cases is on market clearing and relatively stable trends. In considering priorities for NEMS development, subsequent discussion identifies the mid-term modeling capability to be of highest priority.

Long-term energy models projecting beyond 25 years are needed primarily for research and development (R&D) planning and for certain types of assessments, such as the environmental impacts of greenhouse gas emissions. The committee believes that the emphasis of long-term modeling should be relatively simple models and scenarios that incorporate fundamental technical and economic principles. While the distinction between mid-term and long-term approaches is not clear-cut, a key characteristic of long-term models is their ability to accommodate fundamental changes in current trends that are generally not anticipated by mid-term models.

For some types of policy analysis, the use of both short-term and mid-term models (or both mid-term and long-term models) may offer the best approach. For example, an analysis of policy options to deal with future oil market disturbances might draw heavily on short-term models (e.g., those related to the draw-down of strategic petroleum reserves and emergency fuel switching capabilities) in conjunction with mid-term models that estimate the types and levels of future fuel use (including the effects of policy initiatives such as gasoline taxes or energy efficiency incentives). In general, therefore, all three modeling approaches are ultimately required to address the full range of energy policy issues.

General Approaches To Modeling

Two general approaches can be used to construct a model: the comprehensive strategy and the iterative problem-directed strategy. The following comparison is based closely on Initiation of Integration (EPRI, 1977).

The iterative, problem-directed strategy emphasizes in all its phases the interaction between the client and the analysts. The analysis activities assemble information and transfer it to the client in a form that is tailored to the client's problems. The client provides feedback to the analysts on important problems and information needs that can be used to guide the direction of future analysis.

The iterative problem-directed strategy is conducted in three phases. The first phase involves description of the decision alternatives available to the client and identification of information crucial to making the optimal decision. The second phase is analytical. The analyst assembles from the available models, data, and information the combination that is best suited to the client's needs. Where uncertainty is present and is crucial to the comparison of alternatives, a probabilistic formulation of the analysis is structured. The resulting analytical structure is then used, in close collaboration with the client, to generate quantitative projections of key variables bearing on the client's problem and quantitative comparisons of the decision alternatives from the individual perspectives of the client and other concerned parties. The third and final phase of this strategy is the transfer of the analytical output to the client and feedback from the client as to the needs for additional analysis. If the first and second phases have been done in close communication with the client, the third phase is straightforward.

The efficiency and quality of the iterative problem-directed strategy is greatly enhanced if the development of analytical tools is done within a common framework of accounting conventions and variable definitions.

The comprehensive modeling strategy centers on the development of a comprehensive model system covering a wide range of issues of potential concern to clients and providing forecast information of common concern to many clients. The principal product of this strategy is thus the models and particularly the forecast information they generate.

The comprehensive modeling strategy begins with the formulation of an overall analytical framework within which specific modules representing energy supply, demand and other information can be integrated. When the development of these submodels and the overall integrating model are complete, then the submodels are integrated to produce forecasts and analyses of interest to clients. In the comprehensive modeling strategy, the nature of the model is primarily determined by the research process, thus requiring less involvement of the client. This however makes the transfer of the information to the client more difficult.

Given the need to build a national energy modeling system, it seems inevitable that significant elements of the comprehensive modeling strategy will have to be employed. Nonetheless the committee is favorably disposed towards certain elements of the iterative problem-directed strategy, for instance the close involvement of analysts with clients, and the focus on decision problems of importance to those clients.

It follows that the NEMS model builders should be continuously well-informed about the types of policy initiatives and R&D strategies that may be informed by the use of the model and this awareness should shape design and implementation decisions about the NEMS. It also follows that the NEMS should be built as a modular system with several versions of the same module differing in size and complexity. In this way, if detailed analyses are required of only one energy sector, then that particular module can be very detailed while most of the other modules can be relatively simple. In this way, the NEMS as employed in analysis can be kept relatively small and transparent.

The Benefits of Models

The primary advantage of models is that they enable the user to better understand phenomena of interest, based on available information. A model provides a framework in which to analyze different combinations and various kinds of information (e.g., on the magnitude of variables and on cause-and-effect relationships). Such a framework provides insights that are not obtained by analyzing less organized, separate pieces of information. Counterintuitive but valid results from modeling sometimes offer a deeper understanding of the phenomena studied.

Models also force the analyst to be explicit about the assumptions of the analysis. Such assumptions, like those about parameter values or structural relationships, can then be debated or verified by independent observers. Similarly, results of a given model can be replicated from a common set of input values, and differences in results can, in principle, be traced to differences in assumptions. Often, the effort to specify needed model assumptions reveals gaps in essential information (whether in data or in concepts) that the analysis requires.

The well-defined link between assumptions and results means that models provide a convenient way to examine changes in outcomes as inputs change. This is a valuable attribute in policy analysis: sensitivity analysis can reveal the nature of tradeoffs involved in policy interventions and can help to identify an appropriate level of policy intervention.

Models can also help verify results obtained from other forms of analysis. Where the conclusions of different analyses are inconsistent, models can sometimes reveal the reasons for the inconsistency, such as differences in implied assumptions.

The Limitations of Models

While analytical models have come to play an increasingly critical role in policy analysis and planning, the inappropriate use of models, or the failure to understand their limitations, also must be of major concern in developing a NEMS.

Models are simplified descriptions of reality. The simplified descriptions represent those aspects of the real phenomena that the modeler believes to be the most important for the issues under consideration. Any simplification may prove to have omitted significant considerations, which can produce invalid results.

Models impose a fixed structure on an analysis, including the structure that describes the process of change (which is usually based on historical experience). Input assumptions also are often based on past experience. However, future reality may differ in critical ways from the history embodied in the model. In particular, models that fail to incorporate the recognition that the policies they analyze can themselves cause future structural changes may be especially misleading.

Because models produce numerical results, users often tend to attach greater value to such results than are warranted. Often the numbers take on a life of their own, as if they represented reality rather than a highly simplified characterization of reality. Thus, models may lead to a narrowing of the analyst's perspective, rather than to the discerning of new insights.

The aura of reality that attaches to model results often leads users to expect too much, and model builders to promise too much, from this kind of analysis. The desire for realism leads model builders to push toward increasing levels of complexity, placing at risk the simplicity that many times adds most value to a model. Excessively complex models become "black boxes" whose underlying assumptions may no longer be apparent and whose results have implications no longer easy to discern either by the model builders or model users. It is important to remember that model outputs are not "facts" and models are not reality.

In general, several elements affect models' accuracy: (1) the overall structure of the model; (2) the omission of relevant factors that may significantly affect the modeled outcome; (3) the lack of dynamics in the model structure (i.e., assumptions, functional relationships and parameter estimates that are invariant in the model but which actually change in the real world); and (4) the use of data and information of inherently poor quality.

Another serious limitation of most models is that they fail to consider uncertainties explicitly. This issue is especially important for NEMS and is discussed more fully below.

THE MISSION AND FUNCTIONS OF DOE AND EIA

DOE was established in 1977 to manage U.S. energy programs more effectively than its predecessor organizations, and to coordinate national energy policy. The responsibility of the Secretary of Energy is to bring knowledge of energy issues to bear on the formulation of national policy. An important requirement for DOE--and thus for NEMS--is to provide the information and analytical support that the Secretary, a principal client for a NEMS, requires.

Within DOE, the EIA was established and charged with broad responsibility for providing energy-related data and analysis. The EIA was set up to be an independent and objective source of energy-related information, including forecasts of energy trends. By law, the EIA Administrator has independent authority to collect information, conduct analyses, and publish reports (P.L. 95-91). As currently envisioned, the NEMS will be developed and maintained by EIA, both for its own use and to provide services to DOE and others. Thus, EIA would also be a client of the NEMS.

Appendix C provides more detail on the missions and roles of DOE and EIA, including roles as strategic planner, information provider, and energy R&D manager. Each of these roles is described briefly below. All of these functions should influence the design and use of the NEMS.

Strategic Analysis

Strategic planning involves the examination of likely future U.S. energy trends anticipating emerging policy issues; developing options for policy interventions to address these issues; evaluating and comparing the efficacy of such options; and describing and justifying the policy initiatives proposed. Model-supported analysis can be valuable in several of these activities. The energy outlook and emerging energy problems can be clarified by model-generated projections, such as those for alternative future scenarios. Models can be used to explore the future implications of current or altered trends and of the perceived problems.

In strategic analysis, the effects of policy options are simulated and compared to the outcomes of continued current policy. In some instances, the outcomes of alternative policy options can be estimated quantitatively and the options thus made more amenable to political debate.

The NEMS should become the model framework for strategic planning within DOE. It should be capable of generating alternative future scenarios and testing the effects of policy initiatives. For the most part, such strategic analysis would concern the mid-range horizon of 2 to 25 years.

Data Collection and Information Dissemination

The DOE, particularly the EIA, is charged with collecting and reporting information about the national energy system, including production, transport, consumption, and pricing series of major energy forms, and descriptive statistics on the energy production, transportation, and consumption infrastructure. These data supply the building blocks in constructing energy models. The models, in turn, can be used to help assess deficiencies in the existing data. For example, end-use demand models may require data for specific processes or end-use subsectors that are not currently available.

In light of DOE's mission, moreover, energy information has always included the extension of historical data series through energy projections, and models are an important tool in such forecasting. Special studies of anomalies in energy-related data and relationships suggested by data sets (e.g., concerning the delivery capacity of the natural gas supply system, a subject of current interest) also are supported by models.

The NEMS will depend greatly on DOE data, as well as data from other sources, and might be used by EIA as the model framework to generate projections and conduct special studies. To help satisfy the full information mission of EIA, however, the NEMS will need to be augmented by at least a short-term projection model for annual forecasts.

R&D Program Planning

As administrator of a major energy R&D program, DOE must make annual appraisals of its program, evaluate competing R&D projects, and allocate scarce manpower and funds according to some determination of R&D priorities. The traditional approach to allocating R&D resources is to perform a cost-benefit analysis, in which the potential accomplishments of competing projects, adjusted according to their probabilities of success, are compared to their costs. One approach to assessing the potential accomplishment of a new technology is to evaluate its impact on a reference scenario of the future.

Because of the attention to detail required, it is likely that the benefits of specific R&D proposals will have to be evaluated outside of the general NEMS framework (e.g., see the case study of magnetic levitation transportation in Appendix D). Nevertheless, some assumptions used in these analyses can be derived from the general scenarios generated by the NEMS. Prices for competitive fuels, requirements for energy services, and economic and financial variables are examples.

Insofar as a generally applicable reference scenario for the national energy outlook (generated by NEMS) is employed for input to the detailed R&D evaluations, consistency and comparability among such analyses will be enhanced. This use of NEMS could also generate feedback: the more detailed evaluations of the R&D programs may reveal shortcomings in the general model.

In summary, the responsibilities of the DOE and EIA clearly encompass modeling for the three time horizons earlier noted, i.e., short-term (up to roughly 2 years); mid-term (up to about 25 years); and long-term (beyond 25 years). The consistency that needs to be established between short- and mid-term modeling is also recognized as is the link between mid- and long-term modeling capabilities. More specific recommendations about modeling priorities and requirements for the different time horizons are offered later in the report.

Current Modeling Capabilities Within DOE and EIA

From its inception, DOE/EIA has made use of a wide variety of models (EIA, 1990b). They range from comprehensive macro-models of the U.S. economy to very particular models concerning electric utilities' investment decisions for power plants.

Over the years the EIA and the DOE have contributed to energy modeling through early and continued development of large scale energy models and through active participation and funding of the Stanford Energy Modeling Forum (EMF) and the MIT Model Assessment Laboratory. For example, the Project Independence System (PIES) was initially developed in 1974 by the Federal Energy Administration (FEA), and later called the Midterm Energy Forecasting System (MEFS). A number of other modeling systems were developed in the late 1970s and early 1980s by DOE and EIA. The DOE program offices and national laboratories also developed models useful for their particular technology interests. However, during the 1980s modeling activities declined as the pace and priorities of programs at DOE were redefined by the Administration of President Reagan. The EIA, of course, continued to meet its statutory obligations and published its Annual Energy Outlook with the use of its forecasting models. These past energy modeling and data collection efforts addressed to varying degrees the energy supply, demand and conversion sectors and associated economic measures of the U.S. economy; but continuity of efforts at DOE was lost during most of the 1980s until attention to the first National Energy Strategy analysis exercise stimulated a renewed interest at the Department in policy models. A description of some of the current models appears in Appendix E.

The committee doesn't view this collection of models as representing an adequate energy policy modeling system with regard to the requirements set forth in this report. The models have been used primarily to forecast future conditions of energy supply and demand based on alternative assumptions about such external parameters as world oil prices and economic growth rates, and to explore the impacts of such policy initiatives as oil import tariffs and efficiency standards, with reference to a baseline or projection for the future. Several EIA models have been used to obtain external parameter and data estimates, but they are not part of an integrated modeling system. Typically, EIA forecasts are updated annually and serve as a baseline for special forecasting and projects requested in the following year.

One trend in such EIA modeling has been increased modularity and decomposition of the modeling systems. A related trend has been a greater diversity of modeling methods.

These trends have developed as the coverage of systems and the complexity of energy issues have increased (Conti and Shaw, 1988).

EIA's quarterly Short-Term Energy Outlook presents two-year forecasts of energy supply and demand, produced using the Short-Term Integrated Forecasting System (STIFS). This modeling system has been used by EIA since 1979 for short-term forecasting and related analysis. Occasionally a model is developed for a one-time project. The set of models are continually modified as energy markets and the issues addressed change.

The current EIA mid-term modeling system is the Intermediate Future Forecasting System (IFFS), developed in 1982 (Conti and Shaw, 1988). IFFS partitions the energy system into fuel supply, conversion, and end-use sectors and then solves for supply-demand equilibrium by successively and repeatedly invoking the modules that represent these sectors. The model provides forecasts year-by-year and has a forecast horizon of 2010. Fundamental assumptions for this modeling system are world crude oil price and a baseline macroeconomic forecast.

In parallel with EIA activities, DOE's Office of Policy, Planning and Analysis supports an independent integrating model called Fossil2 (AES, 1990; DOE, 1991c). This is the integrating model employed for the recent NES analysis. Fossil2 is a mid-term systems dynamics model that has evolved over a decade. In the recent NES exercise, it was also used to obtain long-term projections out to the year 2030.

In summary, the models available within DOE, especially the EIA models used to make mid-range projections and the Fossil2 model used in policy analysis, already provide significant modeling capability. In the recent analytical effort leading to the NES, however, a number of policy issues proved beyond these models' scope. Thus, the committee finds that the set of DOE and EIA models used in the NES exercise does not constitute an adequate NEMS meeting the requirements for policy analysis described in this report. (This is further elaborated in Appendix B.) The suite of existing models within EIA, however, appears to provide a reasonable starting point for the proposed design of NEMS described in Chapter 3. However, in developing the NEMS, the committee believes the existing models should be modified to address particular high-priority policy issues. Furthermore, Chapter 3 describes the capabilities of some of the models and how they need to be modified.

NEMS IN THE BROAD CONTEXT OF POLICY ANALYSIS

The NEMS to be developed by DOE/EIA must operate in the broader context of an analytical system that for convenience we call the National Energy Analysis System. As illustrated in Figure 2-1, comprehensive models are only one of several tools important for national energy analysis and decision making. Analytic results will derive not only from the models incorporated in NEMS, but also from various data sets, other external models and a wide variety of independent judgments and assumptions. The dashed line in Figure 2-1 indicates that the EIA is involved in all aspects of the analysis process, but its main focus is on models and data sets.

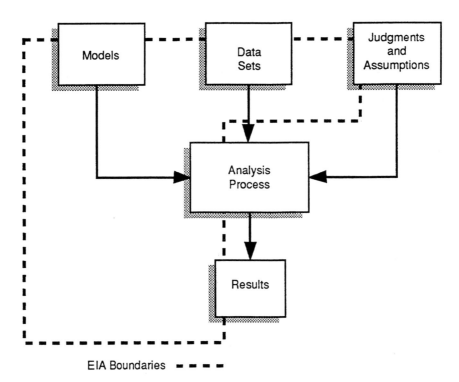

FIGURE 2-1 Scheme of the National Energy Analysis System and the EIA's scope within it.

The development of an appropriate set of models for the NEMS requires a good understanding of the broader context within which these models will be used. Figure 2-2 shows very schematically how the NEMS will interface with parts of DOE, as well as with other public and private organizations in the United States and abroad.

NEMS must establish links both within and outside of DOE to ensure that it remains responsive and able to fulfill its mission. The questions posed by policy makers, and the kinds of information needed for energy policy decisions, moreover, must drive the design of the NEMS. Formal as well as informal processes must be established to ensure that communication between NEMS modelers and their clients is good. The development of the NEMS and associated data bases will be dynamic, evolving in response to changing policy issues and needs. Without effective communication, modelers are unlikely to anticipate and develop the relevant tools and information in a timely way.

Keeping NEMS Outward Looking

Significant data resources and modeling capabilities exist in the private and the nonprofit sectors, and in other state, regional, and federal agencies. To avoid duplication of effort,

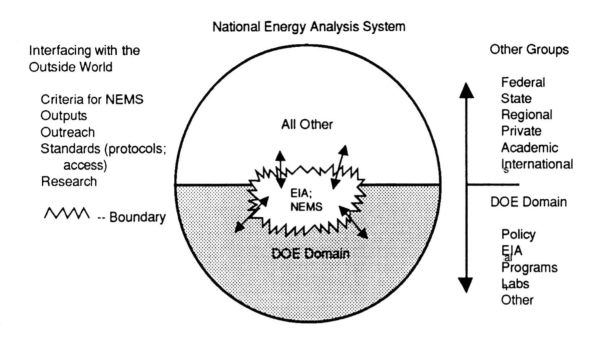

FIGURE 2-2 Scheme of the interface between the NEMS and the National Energy Analysis System.

and to take advantage of the specialized expertise and capabilities of other organizations, the NEMS design should allow outside models to interface with those developed by DOE/EIA. While EIA's use of external models will depend on particular circumstances, it is important to recognize that models outside DOE/EIA constitute a valuable resource that should be considered in developing the NEMS.

Like any modeling system, NEMS will be dependent on its input data and underlying assumptions. NEMS needs valid, current data on resources, technologies, and consumer demands, and sound assumptions about the projection of current trends. By maintaining an outward-looking philosophy, NEMS can draw on the detailed work of outside study groups and detailed sector models, and can help identify and support improvements in key areas.

EIA should identify annually those energy supply, demand, and technology areas that require a new look, as, for example, the impacts of the emerging demand-side management movement in the electric power industry (OTA, 1985). DOE should initiate and participate in broadly based efforts to compile available information on such issues and to increase knowledge about them in the modeling community. The Energy Modeling Forum (EMF) at Stanford University provides an example of how this might be done (Sweeney and Weyant, 1979).

Peer review of models and forecasts can be painstaking and sometimes costly. Yet there is no better way to expose a model's "soft spots" and limits. While the EMF has provided a major service to DOE and other modeling groups by sponsoring such peer reviews, their resources are too limited to ensure comprehensive review and response. Instead, DOE might annually compare selected NEMS results to outside models to identify data and assumptions that lead to differences. This exercise would provide new insights and identify areas requiring better data and more study.

Capabilities Outside DOE

The great diversity and importance of the energy system to many different groups has resulted in a variety of special and general-purpose energy models and tools. The many energy modeling capabilities outside of DOE and EIA include the following:

o Baseline forecasts of the national economy by private econometric firms, such as Data Resources, Inc. (DRI), Wharton Economic Forecasting Associates (WEFA), and public agencies, such as the Congressional Budget Office and Office of Management and Budget, develop basic information on economic growth, inflation, and employment, which helps quantify trends affecting energy demand (e.g., building construction, industrial output, and transportation needs).

o A small number of general-purpose energy modeling systems provide periodic forecasts that broadly consider the total energy market but also have a sectoral or geographic emphasis, for example:

- The Gas Research Institute's annual baseline forecasts for energy, which strongly emphasizes natural gas, particularly new supply areas and technologies

- The California Energy Commission's biannual outlook for energy which emphasizes the environmental, market, and supply concerns of the state.

- The Northwest Power Planning Council's as well as the Electric Power Research Institute's models.

o A great variety of special purpose energy models and studies examine in detail a specific fuel or demand sector, for example:

- The biennial estimates of the U.S. gas resource base by the industry-sponsored Potential Gas Committee

- The assessments of regional electric system reliability by the North American Reliability Council.

o Various general and special-purpose energy models and forecasts are provided by think tanks, consulting firms, nonprofit organizations, and energy companies. Some of these models, data and forecasts may be useful as inputs for some types of NEMS analyses.

o Various international organizations, including the United Nations, the International Energy Agency, and the Organization for Economic Cooperation and Development, provide models and data useful for domestic energy analysis.

o Other government agencies, especially the Environmental Protection Agency with regards to environmental modeling.

Even this brief list shows that considerable energy modeling capability exists outside of the DOE--capability that often covers geographic, technological, or sectoral information in great detail, and which may be useful for NEMS analyses at DOE/EIA.

While there may be strong pressures within DOE and EIA to maintain all control of NEMS development in-house, such an approach could severely limit the value of NEMS. Clearly, NEMS needs its own integrating model that can be efficiently operated and managed by DOE and EIA. However, private organizations and other government agencies provide rich sources of information, input modules, and sectoral and geographic analyses that would be difficult and costly for NEMS to duplicate.

THE NATIONAL ENERGY STRATEGY EXPERIENCE

The previous sections noted two factors that must be considered in specifying the requirements for NEMS: the fundamental mission of DOE must be supported by the modeling capability, and NEMS must relate to external resources and capabilities. Another basis for specifying NEMS requirements is the very recent and extensive analytical experience of preparing the first National Energy Strategy (NES).

Overview of the NES Exercise

The recently completed NES represents a comprehensive analytic effort to support strategic planning (DOE, 1991a,c). The exercise provides a case study of the role of analysis and the subsidiary role of models in supporting DOE's strategic planning.

The goal of the NES was articulated in the early days of the Bush Administration. A comprehensive energy strategy would be developed to establish broad energy objectives and evaluation measures, and a slate of policy initiatives would be proposed to alter the energy future in the direction of these objectives.

The first step in the analysis was to compile a long list of potential initiatives, primarily through a nationwide series of hearings and written submittals. The proposals acquired in this process were described, classified by issue area, and published in a preliminary report, without evaluation, for further comment (DOE, 1990a). The committee commends this broad public consultation process.

Various models were used by DOE in conjunction with the Fossil2 integrating model to generate a reference case projection of the U.S. energy future to the year 2030. Then composite sets of selected policy options were created in interagency consultations. It is not clear to the committee how analysis entered into the selection and assembly of options for the sets. However, once these sets were assembled, various models were used to evaluate the potential impacts of the policy sets as well as combination of those sets on the reference case future. It is the committee's understanding that a final set of policy options were determined under the auspices of the Economic Policy Council and were adopted for implementation by the Bush Administration and proposed to Congress.

During the foregoing process, the DOE Office of Policy, Planning and Analysis often found it necessary to work very rapidly to define a new scenario, run the models, and produce graphical outputs. For such reasons, the NEMS sometimes must be capable of very quick turnaround analysis.

Committee View of the NES Analysis

There is no comprehensive model or group of integrated models existing within DOE, or probably within the federal government, that has the analytical capability to address the mid and long time horizons required by the NES. In the committee's view, DOE's approach of using available models, and off-line supplemental analysis as necessary, was a rational response to the department's need for expedient support of the NES process. DOE's rough integration of modeled and off-line intermediate analyses to calibrate the Fossil2 model has been a useful way to maintain consistent accounting and reporting of results.

The aggregate structure of the models used, however, had significant limitations relative to the analytical results reported by DOE and presented to this committee. It is important for decision makers to appreciate the limited power of the existing analyses used for evaluating policy choices. It would be misleading to assign too much precision to the results of the model runs, or to presume that the models incorporate sufficient information to judge the future impact of policy choices, particularly impacts beyond a decade or two.

In the presentations to this committee, little reference was made to the validation of the models used in the NES analysis. Policy makers should appreciate the important role of the a priori assumptions and simplifications, and the off-line contributions made by the NES modeling subgroups in shaping the scenarios. To a great extent, such assumptions and off-line analyses, rather than the model per se, dictated the results of model runs (e.g., assuming reductions in annual energy use in the industrial sector; NRC, 1991a).

Illustrative Policies

The NES analytical effort has provided a catalog of many energy policy issues currently of interest to government and society at large. Such issues should be an important input to the architectural design of the NEMS.

A list of such issues was provided by the DOE Office of Policy, Planning and Analysis at the request of this committee. The committee believes that an efficient course of action in developing the NEMS would be to arrive, in consultation with interested parties, at a shorter list of the highest priority issues confronting government policy analysts, and to evaluate modeling capabilities to address these issues. The committee believes that a reasonably robust list of high priority issues can be identified. For example, many of the issues in Table 2-1 have been with us for some time, and will continue to be important over the life of many model developments. This will be especially true if a range of parties are all considered rather than those just considered by EIA or DOE to be relevant and important.

Concentrating the available resources on the development of capabilities to deal with specific, high-priority issues will ensure that the NEMS provides near-term rewards to the DOE and that its design is problem-oriented. A problem-focused approach would modify existing energy models to address as quickly as possible the important issues facing DOE and others NEMS clients.

Table 2-1 lists some of the energy-related issues of concern to DOE (Marlay, 1991). These issues, as well as others, can be found in the report on the National Energy Strategy (DOE, 1991a). The committee recognizes that this list is not comprehensive; nonetheless, it is broadly indicative of the types of issues the NEMS should be able to address. The great majority of these issues are addressed best by mid-term models reflecting a time horizon of about 2 to 25 years. This is the time period needed to implement most policies and to observe their effects although there are a number of policies, because of the slow turnover of capital stock, the change in behavior between generations, and the development of some kinds of technology, that require a longer time frame.

To better understand the detailed requirements of a NEMS in its various applications, the committee undertook three case studies of energy policy analysis, drawn from Table 2-1. These hypothetical case studies, which are described in Appendix D, consider in more detail how the NEMS should be able to be used as a tool in analyzing a broad range of policies. The case studies cover issues in energy demand, energy supply, and R&D program planning.

Several insights were suggested by these case studies and related analysis by the committee. These insights can be summarized as follows:

o The NEMS can provide useful policy guidance on a wide range of issues, many of which have not been addressed by national energy modeling in the past (e.g., various types of end-use efficiency incentives).

o To facilitate this process the NEMS models must have two highly desirable attributes:

  - <u>Simplicity</u>. The level of detail must be limited to that strictly necessary to address the key issues.

  - <u>Transparency</u>. The model should not be a "black box;" rather, the relationships among the variables, parameters, equations, and projected trends must be stated clearly and be well understood by model users. As discussed in what follows, transparency will be aided by good documentation, reduced-form models, outreach

TABLE 2-1 DOE Policy Issues

Effects of R&D (including uncertainty)
    Supply (Enhanced Oil Recovery)
    Demand (magnetic levitation trains, highspeed rail)
    Alternative fuels (fuel blends; expanded fleets)

Effects of Deregulation
    Natural gas (pipeline certification)
    Electricity (lift size cap on PURPA)

Effects of Taxes and Price Changes and Volatility
    Gasoline ($1 tax, phased in over 5 years)
    Price spike

Effects of Investment on Production Incentives
    Investment tax credit
    Renewable energy production credit

Effects of Command/Control Regulation
    Corporate Average Fuel Economy (CAFE) standards
    (34 or 40 mpg by 2000)
    Lighting standards (bulbs, ballasts)

Effects of Information (efficiency labeling)
    Fuel economy information
    Expanded appliance labeling
    Efficiency labeling (light bulbs, fluorescent light tubes)

Effects of Resource Estimates/Recovery Costs
    Outer Continental Shelf
    Arctic National Wildlife Refuge
    Natural gas (cheap vs high cost)

Source: Marlay (1991).

activities, and the development of NEMS for use on personal computers (NRC, 1991c).

o Simplified versions of the models relating their inputs to outputs (reduced-form modules) can be helpful in increasing the transparency of the full modules. Such simplified representations may be obtained by running the complete representation of a given module many times to generate data; then statistical methods can be used to estimate a relatively simple "response surface" relationship between the input and output variables. Judgments then can be made about the reasonableness of the estimated parameters of the reduced-form module to characterize the validity and robustness of the results.

o Not all policy analysis will be done directly within the NEMS; some will require ancillary off-line simulations. However, the NEMS can provide a consistent framework or baseline to discipline and support these off-line analyses, leading to greater consistency in final results.

o The NEMS will require more disaggregation and greater data related to energy demand and supply. These needs are elaborated later in the report.

o The NEMS should provide better simulations of human behavior: how energy users and producers are likely to respond to economic and other stimuli, including how incentives (e.g., consumer rebates for more efficient cars, guaranteed-price contracts to energy producers) are likely to affect energy-related behavior. The major issue is the extent to which consumers and producers make economically optimal decisions with regard to energy.

o Focusing on particular policy issues can help to clarify where specific models need to be improved.

DIRECTIONS IN ENERGY MODELING

The science of mathematical modeling to assist in policy analysis and energy choices is very dynamic. Better techniques are continually being developed to analyze existing and new data. The modular modeling systems now becoming available can more quickly accommodate these better techniques (see e.g., NRC, 1991c).

Fresh energy paradigms are emerging that are more insightful and more comprehensive. Examples include the incorporation of environmental and social costs associated with the production and use of energy (Hubbard, 1991); recognition of regulatory and fiscal incentives and disincentives and the subsidies that support various energy systems (Stobaugh and Yergin, 1979); and recognition of the inadequate treatment of activities causing resource depletion and environmental degradation in measures of the gross national product. Inherent in a modular design for the NEMS as discussed below and in Chapter 3, is the ability to assimilate changes as they mature.

General Trends

A number of more detailed considerations for the design of NEMS emerge based on general trends in modeling:

Integration of engineering detail and economic decision-making algorithms. Particularly for the analysis of energy consumption and demand, engineering detail and economic decision-making algorithms are being better integrated, for several reasons (McFadden, 1983; Dubin, 1985; Dubin and McFadden, 1984). First, technological and econometric approaches have both been recognized as providing information and insight about energy demand behavior; second, greater technical information has become available from studies conducted over the past 15 years; third, it has become apparent that consumers may not follow simple rules, based on engineering-economic criteria, in purchasing energy-using capital equipment; and finally analysis of policy initiatives that increasingly affect consumer choices (e.g., building and appliance standards and utility efficiency programs) require more technical and economic information.

Such new modeling integration is also incorporating the concept of self-consistent or rational expectations. Many energy decisions, whether by households or businesses, depend on expectations about future energy prices. All energy models therefore incorporate, either implicitly or explicitly, the formation of price expectations. In contrast to most models of the past, the current direction in energy models is to model price expectations so that they are consistent with the future price realizations.[1]

Modularity of system design and construction of integration algorithms. System design is becoming more modular to enable a large number of modelers to work separately on particular aspects of large systems. This specialization of labor entails that modules be designed with specific input and output requirements so that the modules can be linked together to derive a solution. Efficient algorithms also must be designed for linking the modules.

More explicit representation of uncertainty. There is greater recognition of the uncertainties associated with energy modeling, uncertainties that result from the long time horizon required to study many energy issues, and the many unknown but important variables and behavioral parameters that affect energy trends. This recognition has led to the use of an array of analytical methods that range from wider use of scenario analysis to portray alternative futures, to the use of more formal techniques to represent the probability distributions of outcome variables of interest.

Use of contingent strategies. As uncertainty is introduced into energy models, it follows that the modeled energy system may diverge from the most expected or most desirable path

---

[1] See Chapter 3. Myopic expectations: individuals assume future prices will remain the same as current prices; extrapolative expectations: individuals extrapolate recent trends in prices; rational expectations: individuals forecast prices consistent with a model. The price expectations need not be correct but rational expectations are the only self-consistent way to model expectations.

(NPPC, 1991). If this occurred in the real world, then presumably policies would be adopted to redirect the energy system to a path believed more desirable. Thus, for example, corporate average fuel economy (CAFE) standards would probably not have been enacted if not for the rapid oil price increase in 1973-74. To replicate this real-world behavior, therefore, an energy model should incorporate a way to redirect the solution to the desirable path (as via new policies or options). Thus, consideration of uncertainty in energy models implies the need to model contingent strategies.

Transportability of codes due to the broad availability of computers. If the system were implemented for widely available computers (personal computers), it could be made broadly available to other analysts. Such availability would increase the numbers who could understand, use, and critique the system. Unless EIA implements NEMS on personal computers its models will remain accessible only to very few users. Such limitations on accessibility would needlessly reduce the credibility of the system and limit EIA in its ability to receive valuable advice and criticism from a wide array of analysts outside the DOE.

The current state of personal computing technology allows ample computing power and memory for even complex modeling. But it is possible (although the committee believes, not probable) that if NEMS were implemented on personal computers, the NEMS models' content would be sacrificed. If that were the case then EIA might wish to configure NEMS to be usable on workstations, rather than on personal computers. Configuring it for workstations would reduce the breadth of use by analysts outside of DOE and would thereby reduce the national benefits available from the system. However, such a configuration would make the modeling system more accessible than if it could only be used on mainframe computers, as is currently the case.

The committee therefore recommends that NEMS be developed to run on personal computers unless such a restriction would by necessity significantly sacrifice the NEMS' content. In that case, development for use on workstations would be regrettable, although possibly reasonable. In no case should the system be developed so that it could be readily used only on mainframe computers.

Multi-attribute utility functions. Energy policy decisions often have conflicting economic, environmental, and national security implications (e.g., energy developments that harm the environment or increase costs). The theory of multi-attribute utility analysis has been extensively developed to deal with tradeoffs and identify optimal decisions to be made by a single decision maker (Keeney, 1988). In energy policy decisions, however, there usually is no single decision maker, and policy choices have effects on different groups with different goals. As a means to better understand the sources of potential conflict and opportunities for compromise, approaches for synthesizing diverse model output measures are being sought, as are methods to reflect the values of different groups likely to be affected by energy policy decisions.

Quantification of Uncertainty

The importance of representing uncertainty in energy policy models is especially emphasized by the committee. This section discusses the difficulties such uncertainty poses for energy modeling, and approaches that might be incorporated in the NEMS.

Because policy analysis models are necessarily built with both incomplete theoretical foundations and incomplete empirical data, both the modelers and users of results should be concerned about the uncertainties inherent in the results of energy policy modeling. Such uncertainties may arise from a number of sources including data errors, exogenous forecast errors, future technology characteristics, specification of dependencies, estimation error, and modeling behavior. A more detailed treatment of these and related issues may be found in Morgan and Henrion (1990).

Most contemporary models of the energy system are deterministic. They require specific input variables, and in theory, a level of detail that almost always greatly exceeds the precision with which real-world variables are measured. For example, the use of energy by the entire U.S. household population is described in DOE models based on the survey reports of only a few thousand respondents with virtually no physical measurements to confirm the data.

As noted earlier, model structures or algorithms represent relationships that are simplifications of historically observed or assumed relationships among the variables. For example, energy-related consumer decisions typically are represented as economic choices among available options exclusive of other needs or desires that may be competing for the same investment.

The increasing range of possible outcomes that proceeds from alternative decisions as the future unfolds is usually ignored. For example, many energy models assume a fixed-price track for the world oil market, that continues to guide future model decisions, even after intermediate results of the model run, such as those relating to oil demand, have become inconsistent with that track.

These and other simplifications introduce large uncertainties in the results presented to decision makers as a basis for action. Uncertainty is inherent in the nature of models and cannot be eliminated. Nor should it be ignored.

It is important to deal explicitly with uncertainty for a number of reasons. First, one desires actions that are robust across a variety of possible situations. An explicit treatment of forecast variability allows an evaluation not only of the success of a policy in some "expected" or most likely future, but also its success in, or sensitivity to, a range of alternative futures.

With the recognition of uncertainty comes a better appreciation of risk. If risks can be characterized and quantified, an appropriate risk management strategy can be undertaken. For example, by quantifying the likely potential variability in electricity demand forecasts, the Northwest Power Planning Council (NPPC) has been able to consider the robustness of their decisions across possible levels of regional economic activity. Because future loads are uncertain, the NPPC developed a range of estimates plus analytical models to evaluate the impacts of uncertainties on the distribution of future power systems costs. Both expected costs and uncertainty in costs were estimated for alternative resource strategies using Monte Carlo methods to combine hundreds of future scenarios into probablistic statements about key parameters of system performance. As a result of such analyses, in one case NPPC chose to reduce risk in the face of uncertain future demand levels by investing in higher-cost resources in the near future (Northwest Power Planning Council, 1991). Hirst and Schweitzer (1990) give other examples of electric utilities' use of strategic models to accommodate for uncertainty and risk in their long-range planning.

New applications of uncertainty analysis to advanced energy technologies and R&D planning also are emerging that may be of significance for mid-term and long-term energy modeling. Diewekar and Rubin (1991), for example, have developed probabilistic models that allow DOE to estimate the performance and cost uncertainties for advanced power generation systems. Illustrative results, based in part on the technical judgments of DOE experts, show a much broader range of cost and performance uncertainty than reflected in current deterministic estimates of the type now used in energy policy models (Frey and Rubin, 1991). To the extent that policy analysis results depend strongly on assumptions about future energy technologies (as often they do), failure to adequately incorporate uncertainties could lead to very misleading results.

For long-term analysis consideration of uncertainties is especially critical. In their analysis of global warming, for example, Manne and Richels (1990) determine an optimal energy path for the United States for the next 20 years given three possible scenarios for U.S. energy development starting in 2010. The scenario choice in 2010 is determined by a judgment about the severity of future global warming. Again, the explicit treatment of uncertainty engenders considerations about the robustness of potential policy options across a range of future situations.

In all cases, attempts to explicitly quantify current ignorance may also help direct the collection of additional data, or the revision of model structures, to reduce the most critical forms of uncertainty for future analyses.

The NEMS can be designed to accommodate various methods for explicitly treating uncertainty in energy models, including sensitivity analysis, closed-form statistical approaches, Monte Carlo methods, and alternate model formulations. A modular structure for NEMS can facilitate uncertainty analysis, first, by allowing many of the full modules to be replaced by corresponding reduced-form modules, thus considerably reducing the run time, and secondly, by allowing the use of alternative module structures for a particular segment of the energy system (for further discussion of uncertainty see Cohen, 1986; Hausman, 1981; Heyde and Cohen, 1985; Hodges, 1987; Leamer, 1983; Malliaris and Brock, 1982; and Sims, 1982, 1984).

Long-Term Forecasting

While the focus of the committee's attention has been modeling for the medium-term, there is also a clear need to consider longer time periods. A mid-term projection horizon of roughly two to 25 years contemplates less than a complete change in the current generation of adult decision makers, which suggests some continuity of behavior in household customs and business practices. Similarly, much of the energy infrastructure of 25 years from now already exists or will result from currently observed trends. Important new technologies that are not evident in some form today are unlikely to penetrate the marketplace significantly within 25 years; and structural changes in the makeup of the economy are less likely to diverge from current trends within that time.

None of these assertions can be made with the same degree of confidence for a longer-term projection horizon. As a result, the simple extension of a mid-term energy model for an additional decade or two might suggest a degree of certainty out of proportion to the confidence that can generally be placed in the method. In fact, the "inertia" in the model's relationships, rather than adding information, may give rise to misleading results. For long-term projections, the past loses relevance as a guide.

Since long-range projections are needed for some types of policy analysis, an entirely different method may be required. The notion of predictability, central to the conventional meaning of projections and forecasts, should be de-emphasized, while that of developmental constructs or working hypotheses should be stressed.

Long-term modeling methods should be based on a critical selection of those relationships and trends that are most likely to persist (e.g., certain demographic trends). Judgments about the long-term direction of important generic factors, such as the efficiency of energy conversion, should be made based on fundamental physical or economic principles, rather than by extrapolating current experience. Factors that are fixed in the short- or medium-term must become variables in the long term, including social and political factors that shape the economy today.

In developing the long-term capabilities of the NEMS, therefore, it would be well to begin with the broadest notions about alternative futures. For instance, the rapid development and deployment of information and communications technology over the next decade or two may make it possible for employees in many service industries to perform most of their work at home, thus eliminating a large fraction of automobile usage. The South Coast Air Quality Management District's 20-year master plan for the Los Angeles Basin specifies air quality standards that cannot be met by conventional internal combustion engines or their foreseeable successors. If followed, this 20-year plan might lead to a future dominated by electric or alternative fuel vehicles, with profound implications for gasoline usage, and economic and environmental outcomes.

Because over the long term there may be significant changes in structural relations (i.e., in those relationships often treated as invariant to medium forecasting), long-term models must be able to anticipate structural changes. In these cases, the parameters and coefficients that define the various relationships should be obtained from basic technological information, exogenous models, or subject matter knowledge.

The attributes of simplicity and transparency that are desirable characteristics for all models are especially important for long-term models. In addition, a third characteristic applies exclusively to long-term models:

o **Judgment-Based.** Because fundamental structural changes are likely over the long term, greater emphasis needs to be placed on judgments and reasoning about basic assumptions rather than on extrapolation of current trends or the use of mathematical relationships applicable for mid-term modeling.

The main ingredients of long-term modeling are core assumptions and judgments, representing the modeler's outlook on the context in which projected trends will develop. If the core assumptions fail to capture the reality of the future context, the modeling used will make little difference to the quality or utility of the results. The results from long-term models should not be interpreted as predictions of future events but rather as tools to provide insights or indications about where the basic assumptions might lead.

A common practice in long-term modeling is to conduct scenario analyses to assess the impacts and implications of changes in regulations, laws, or the assumptions themselves. Useful information can be derived from such scenarios. For example, important issues that recur in many scenarios can be identified. Extreme situations and the most threatening or beneficial consequences of potential trends can be explored. Gross requirements for energy resources and cumulative consumption quantities can be estimated and compared to resource availability. Pollutant loadings can be calculated and compared to the environment's assimilative capacity. To some extent, the longer-term impacts of proposed policy initiatives can be described. The scenarios, however, should not be treated as quantitatively as mid-range projections. It is advisable that some scenarios also be free of the imposition of "plausibility" or "sanity" checks. <u>In thinking about longer-term models, it is useful to remember that conditions that today seem implausible could be normal in the distant future.</u>

Different degrees of modeling detail are necessary in long-term models compared to the mid-term model; however, these differences depend on the peculiarities of the sector being modelled. Considerable thought and analysis is necessary before the modeler can choose rationally the types of methods used for long-term forecasting. Modelers disagree on the extent of disaggregation that is useful with such long-term models. Those modelers emphasizing the extreme uncertainty of such models suggest simple, highly aggregated models. Others feel that disaggregated models, especially on the demand side, can be quite useful and illuminating. Most modelers agree, though, that these models will be different than those use for the mid term.

The committee heard a number of presentations on efforts outside of EIA that address long-term energy issues (see Appendix E). For example, Manne and Richels (1990) and Edmonds (1990) are modeling on a time scale of 100 years in studying the global carbon cycle and its connection to the energy system. Others discuss option theory and hedging as strategies to guard against an uncertain future. And, of course, there is much work directed toward modeling long-term climate and environmental change. DOE and EIA have a rich assortment of activities to look out to in embarking on the development of a long-range modeling capability.

While the committee believes that the highest priority must be given to the development of a NEMS mid-term modeling, EIA also should devote some resources now to studying activities on long-range modeling outside DOE/EIA. Based on this review and the needs of the NES process, DOE/EIA should devote additional resources to the development or acquisition of appropriate long-range modeling capability, consistent with the discussion above.

NEMS REQUIREMENTS

Based on the previous considerations, the committee recommends that the requirements for a viable and useful NEMS include the following:

1. **Output measures of concern to decision makers:** NEMS output should including key economic, environmental, and national security measures relevant to policy decisions (as outlined below), plus income and regional distribution characterizations as appropriate.

- **Economic measures:** Key economic outcomes include measures of changes in consumers' and producers' surpluses from a change in the supply curve of energy as a means of determining the societal cost savings associated with such a change (Willig, 1976), gross national product (GNP), and federal budget effects. Since the value of energy services is an important issue in contemporary energy policy discussions, an attempt should be made to report this variable as well.

- **Environmental measures:** Ideally, the environmental characteristics modeled should include both flows and stocks of major identified pollutants and, to the extent possible, measures of their economic or physical effects. Measures of environmental effects are very imprecise at this time. The committee believes that initially NEMS should estimate the total emissions of primary pollutants and related measures of environmental impact (see Table 2-2). Later, secondary or indirect impacts should also be quantified (e.g., land use needs for biomass). Thus, NEMS development should be capable of supporting models of environmental effects developed by other groups or agencies, such as the Environmental Protection Agency. Ultimately it may be desirable to incorporate simple models of environmental effects directly into the NEMS.

- **Energy Security measures:** Energy security is a derived characteristic. NEMS should attempt to measure the probabilistic effects on the economy and the environment of various potential disruptions of the energy system. The security concern that has received the greatest attention is that associated with disruption of the oil market. One proxy variable that was proposed in the NES to measure the vulnerability of the U.S. economy to oil disruption is the aggregate cost of oil relative to GNP. The larger the ratio of oil cost to GNP, the more vulnerable is the economy to oil price

TABLE 2-2 Environmental Issues Relevant to NEMS

General Concern	Problem Area	Primary Measures
Air Pollution	Global warming	CO_2, CH_4
	Acid rain	SO_2, NO_x
	Photochemical smog	VOC, NO_x
	Air toxics	Metals, organics
Water Pollution	Oil spills	Crude, products
Waste Disposal	Nuclear wastes	High-level waste
	Non-nuclear wastes	Ash, sludge, other
Natural Resources	Land use	Acreage
	Water use	Consumption
Health and Safety	Nuclear accidents	(a)
	Other energy-related accidents	(a)
	Electromagnetic fields (60Hz)	(a)

(a) To be defined in later studies.

or supply disruptions. Deeper understanding of the effects of oil disruptions is required, and could be gained from more focused studies relating to the role of oil in the U.S. economy and the structure of the world oil trade. Once these studies are completed, a better proxy measure of oil vulnerability probably can be designed and incorporated in the NEMS.

o **Regional and international measures:** In general, output from NEMS will be required for the United States as a whole. Many issues, however, will have regional or international implications. Environmental impacts and transfers of wealth resulting from U.S. policy initiatives are only two examples. Most energy production and consumption patterns in the United States vary significantly by region. From the outset the NEMS will need some regional disaggregation of results (e.g., for some economic and environmental impacts). The specific needs will be defined by the problem-oriented focus recommended below. In addition, NEMS will have to be sensitive to international issues that influence U.S. energy modeling (e.g., world oil price) and that are affected by U.S. energy policy (e.g., global warming).

o **Income distribution measures:** In certain cases, energy policy initiatives may have different effects on classes of people with different incomes. While the first implementation of NEMS presumably will not disaggregate results by income distribution, it will be desirable that it do so in future analyses for some policy initiatives.[2]

The committee also believes it would be appropriate to pursue research on approaches for summarizing the values or tradeoffs for different groups likely to be affected by energy policy decisions.

2. Use of the iterative problem-directed approach. Given the diversity of energy policy issues of concern to DOE/EIA the development of the NEMS will have to incorporate some elements of the comprehensive modeling approach. Nonetheless, it is the opinion of the committee that NEMS model construction should be driven to the extent possible by an iterative problem-directed approach. Thus, the model builders should be continually aware of the types of policy initiatives that are to be analyzed, and this awareness should determine the priority design and implementation decisions for the NEMS.

3. Use of a modular structure. The design of NEMS should be modular to facilitate its use, and, to enable model builders to work on particular aspects of the model independently. This specialization of labor requires that modules be designed with specific input and output requirements so that all modules subsequently can be linked together to derive an equilibrium solution. It will also be necessary to maintain overall flexibility and to design efficient algorithms to link the modules.

4. Integration of engineering and economic approaches. NEMS models should strive to project actual producer and consumer behavior as well as explore technically efficient outcomes. Integrating engineering and economic decision-making algorithms would permit the NEMS to better project actual consumer and producer responses to changing events, to existing regulations, and to proposed policies. Efforts should also be made to incorporate alternative models of expectation formation. Many energy decisions by households and businesses depend on expectations about future energy prices. To the extent possible, the NEMS should be designed to eventually incorporate alternative models of price expectations, namely, myopic, extrapolative, and rational.

5. Incorporation of uncertainty analysis. There are several methods for accommodating uncertainty in energy models, including scenario analysis, parameter variations, alternative model structures, closed-form statistical approaches, and Monte Carlo methods. The committee recommends that NEMS be designed to accommodate the explicit treatment and estimation of uncertainty using a variety of methods.

In developing and using the NEMS, uncertainty should be explicitly addressed in two ways. Initially, the existence of uncertainty in analytical results should be brought to the attention of decision makers in every analytical report. Uncertainty also should be made

---

[2]In Chapter 3, the architecture proposed for NEMS indicates that certain analyses, such as calculating the economic distributional impacts, would be analyzed with models separate from NEMS (see Figures 3-1 and 3-4).

explicit by describing the limitations of data and algorithms in the most accurate way possible.

6. Focus initial development on the mid-term. To address the full range of energy policy issues confronting DOE, model results generally are desired for three time horizons: (1) the short-term, roughly 2 years; (2) the medium-term up to about 25 years; and (3) the long-term, beyond 25 years. Although the NEMS is intended to cover all time periods, the committee agrees that its initial focus should be on the medium term. However, because the impacts of many policy initiatives, and many of the R&D questions of concern to DOE, will involve evaluations beyond a 25-year time horizon, DOE should ultimately develop adequate long-term modeling capability.

The problems of long-term modeling are unique. While there is a body of experience available, original research will likely be required to develop a meaningful modeling capability within the NEMS to evaluate the long-term consequences of policy options. The treatment of uncertainty and of alternative outcomes become very significant in such evaluations.

It is recommended that the DOE/EIA create a group to develop a long-term modeling capability. In addition to relating this capability to the medium-term modeling system in a consistent fashion, this group should incorporate the expertise of external experts and groups who already focus on the very long term in energy modeling.

Short-term modeling already is used by EIA in preparing its short-term energy projections. Because the focus of short-term models is on economic cycles, base-year anomalies, supply disruptions, and monthly or quarterly outputs, the methods employed are inconsistent with longer-term model structures. Development of short-term capability for the NEMS framework will thus need to focus on compatibility and consistency with the mid-range framework to the greatest extent possible. At the outset, however, the short-term capability will likely continue to be separate from the mid-range modeling system.

7. Use of widely available computers. To maximize the likelihood that NEMS will be usable by the many different parties at interest in the national energy debate, the NEMS should be constructed to run on one or more widely available hardware configurations (e.g., a "personal" computer or a more powerful workstation unless this would significantly restrict the NEMS content). In that case development for workstations would be regrettable but preferable to mainframe computers. The portability of the NEMS code across different computer operating systems also will require attention in this effort.

8. Outreach to the broader energy analysis community. The NEMS should be designed so that it invites interfacing with existing or future models outside of the NEMS. A modular design should facilitate such outreach.

9. Interaction with outside peer groups. Critical reviews from outside peers, participation in workshops, oversight committees and subjecting the NEMS development work to open interaction will all lead to higher standards and greater credibility of the NEMS.

10. Transparency of results. The transparency of the NEMS model should enable users other than DOE/EIA modelers to use parts or all of the NEMS independently.

11. Fast turnaround. A version of the NEMS should be designed to operate in reduced form to facilitate repetitive runs or simplified analytical efforts. Rapid response should be facilitated by reduced-form modeling and abbreviated administrative procedures (in contrast to the detailed procedures currently in place for EIA models).

12. Quality control. In all aspects of NEMS development, procedures for maintaining high levels of quality control will be required. These procedures will include verification and documentation of coding, data and model validity, as appropriate. Adequate flexibility, however, must be preserved to ensure that NEMS development and use does not become stifled by overly elaborate procedures and specifications.

The suggested structure and architecture for the NEMS that can meet these requirements are outlined in the following chapter.

FINDINGS AND RECOMMENDATIONS

The primary committee findings that emerge from this chapter are the following:

- **The set of EIA models reviewed by the committee at the beginning of this study constitutes a reasonable starting point for developing a National Energy Modeling System. However, considerable development will be needed to attain a modeling sysem satisfying the requirements outlined in this report.**

- **The NEMS program, once established, should complement, interact with and draw upon analyses from a variety of other public and private groups that contribute to policy analysis.**

- **Successful development of a NEMS will require the Secretary of Energy and the EIA Administrator to establish and foster an organizational environment that is outward-looking and ensures greater intellectual and institutional commitment to its development and maintenance.**

Although the NEMS needs to satisfy a number of requirements, the committee especially recommends the following:

- **The NEMS should be designed to estimate the economic, environmental, and security implications of alternative energy policies.**

- **The NEMS architecture should be modular and should provide for quick turnaround applications.**

- **The NEMS should be designed to allow analysts to incorporate uncertainty explicitly.**

o **The primary focus of NEMS development should be on capabilities that address the mid-term time horizon of up to about 25 years.**

o **A problem-focused approach should guide the development of all NEMS capabilities, to the extent possible.**

Furthermore, with regard to long-term models, the committee recommends:

o **The EIA should create a group to develop a long-term modeling and analysis capability.**

3

NEMS ARCHITECTURE

OVERVIEW

Chapter 2 described expected uses and desirable inputs and outputs of a National Energy Modeling System (NEMS). In this chapter a general architecture is proposed for the NEMS.

The proposed NEMS should be designed for simulations and analysis relating to the mid-term time horizon, up to about 25 years in the future. While such a modeling system could be used in principle for both shorter and longer term modeling, these kinds of modeling typically have special analytical requirements. For this reason, the committee recommends that EIA not attempt to use a single model for the three kinds of analysis (that is, for short-term analysis, that of up to about 2 years), medium-term analysis, up to roughly 25 years), and long-term analysis, beyond 25 years). In particular, the Department of Energy (DOE) should not select the mid-term model to be used based on the desire to carry out long-term analysis within the same framework.

For short-term modeling, it is usually important to consider supply and demand fluctuations over the course of an annual cycle. Effects of random events--severe weather, strikes, international hostilities--may be of particular interest. Considering adjustments to projected supply and demand disequilibria (based on forecasted starting point derived from the mid-term model) may also be important, for example, when modeling our vulnerability to energy supply disruptions. This report will not consider in greater depth modeling for the short-term time horizons.

For long-term modeling, modeling structures should be much simpler, because in projecting the longer term there are many more unknowns. Long-term modeling systems stress uncertainty about possible outcomes and allow the uncertainty itself to be easily examined. Issues of interest for long-term energy modeling include fuel substitution, physical limitations of natural resources, technological evolution, lifestyle changes, and effects of population growth. Basic concepts for long-term analysis were discussed briefly in Chapter 2, and these also will not be pursued further here.

The proposed modeling system for the medium term would consist of modules linked together in a larger analytical system. It should be possible to run these models separately, all together, or in combinations, depending on the analytical needs.

This modeling system should be designed primarily to simulate or project energy futures based on assumptions about policies and other driving variables. NEMS should also be useful for some kinds of planning and optimization. In optimization, possible future patterns are compared to objectives encoded in the model. For example, patterns of energy use the modeler deems optimal can be compared with observations about actual rates of energy use and production under various conditions. In what follows, it is assumed that NEMS models will be used mostly for simulations rather than optimizations.

NEMS should be structured to project supply and demand equilibria in U.S. energy markets. Prices, which guide the energy system toward a supply-and-demand equilibrium, should be explicitly accommodated in both supply and demand modules (for discussion of general equilibrium analysis, see Ballard and Goulder, 1985; Roger and Goulder, 1984). The modeling system should be able to analyze the impacts of policy options and other variables on the economy, the environment, and the security of energy supplies. Existing regulatory impacts should be represented in the individual modules and the system should allow analysis of contemplated regulatory regimes. In addition, supply and demand modules should allow the examination of non-policy factors (e.g., demographic trends) that, in addition to prices, shape supply and demand.

Individual modules would include energy supply and demand models and an interindustry economic growth model brought together in one "integrating system," and auxiliary models that would generally use outputs from the integrating system once its solution were found. Figure 3-1 provides a simple illustration of the proposed NEMS architecture.

The interindustry growth model would take prices of energy and supply and demand-side investments as inputs and project the economic activities resulting from those inputs. Each energy demand model in the integrating system would take as inputs the prices of energy (P) and levels of economic activity (A), and would produce as outputs the quantities of energy utilized (Q) and the demand-side investments associated with that energy use (I). The integrating system's supply and conversion modules would take as inputs the quantities of energy demanded in this system and provide as outputs the prices at which those quantities of energy would be made available. In addition, the supply and conversion models would provide the levels of investments associated with those energy quantities.

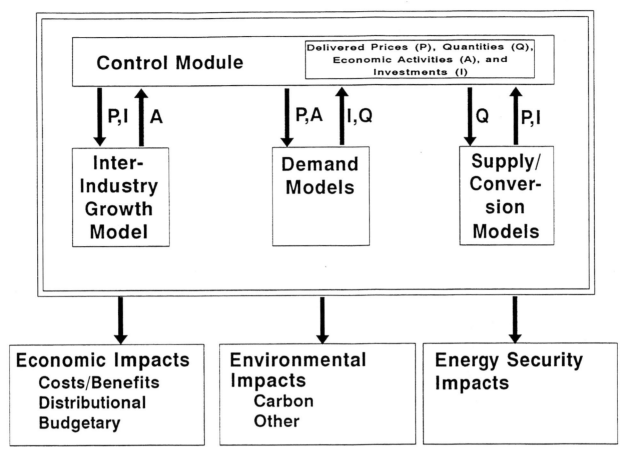

FIGURE 3-1 Simple representation of proposed NEMS architecture.

The models within the integrating system would interact with one another through common data files, which would provide information to models and accept it from them, as appropriate. The modules would run sequentially under the guidance of a control module that would seek convergence to an equilibrium. The control module would take as outputs from the other modules values for prices, energy quantities, and investments, and would store them in the common data files. These values would then be passed by the control module back to the individual modules for the next iteration. The control module would include criteria to determine when the system had come sufficiently close to equilibrium.

The integrating system would be run iteratively until a solution was found. For any iteration, the integrating system would begin with an initial set of prices and investments over the entire time horizon and would predict demand-side investments and energy quantities demanded over that horizon. These values for energy demand would then be passed to the supply and conversion models, which would in turn provide prices and investments as outputs, which would likely not be consistent with the assumed initial set. The control module would include an algorithm to successively adjust the trajectory of prices and investments to converge toward an equilibrium. Equilibrium would be obtained when a fixed point was found, that is, a set of price and investment values such that input and output values were identical for each fuel, region, and time.

Once an equilibrium solution is found for the integrating system, then outputs from each of its modules could be used in the three auxiliary models, those examining the economic, environmental, and energy security impacts of the given equilibrium.

The rationale and for more detail about this proposed NEMS architecture are provided below.

MODULAR ARCHITECTURE

The proposed NEMS system is described as modular. Modularity in modeling, however, is a relative concept: all modeling systems can be viewed as modular in that equations can be removed and other equations substituted and the model can be run. The modular system intended here refers to a group of modules that, each taking the limited set of inputs from the control module, could be run separately from one another. As the committee envisions it, the defining characteristic of a modular system is the ability to run the modules within a system separately. These modules should have common data files, from which they take inputs to perform calculations and to which their outputs flow. The system should be able to run any individual module without any of the other modules being loaded into the computer. A modular system would allow users to replace modules with alternative versions conferring great flexibility to address highly varied analysis requirements.

To implement such a modular system, the EIA must clearly define what data will be passed among the various modules. All such data would be passed through common data files; none would go directly between the component modules, except between the control module and others. EIA must carefully define input and output variables so that they have the same precise definition in all of the modules. Once attention is paid to the interfaces between the modules and the common data files, the internal structure (and possibly even the software) of any module can be independent of any other.

Modularity is not a new concept to EIA modeling, as was noted in Chapter 2. Some modularity characterized even the earliest modeling systems of the EIA and Federal Energy Administration. For example, the Project Independence Evaluation System (PIES) and International Energy Evaluation System (IEES) permitted different demand models to be used to calibrate a constant elasticity demand representation and different oil supply curves to be used in the linear programming representation. The current EIA integrated model--the Intermediate Future Forecasting System (IFFS)--is a modular system.

The NEMS system that the committee proposes would be a logical extension of the IFFS. There would be no need to develop the NEMS from scratch. Initially it could rely strongly on existing models, adapting them to the new configuration.

The EIA now stresses modularity as an important goal. EIA papers assessing existing and desired modeling systems have emphasized the importance of the modular approach (EIA, 1990a). The committee strongly endorses this emphasis.

Advantages of Modular Systems

Modular architecture facilitates the decentralized development and maintenance of modeling system components, the modules (Hogan and Weyant, 1983; Cowing and MacFadden, 1984). Thus, once NEMS model inputs, outputs, and module interfaces are well defined, development of NEMS modules could proceed in a decentralized manner under EIA supervision.

The ability to substitute one component for another in a modular system allows, among other things, running any combination of modules without running the others (e.g., by using "null modules," that do nothing and do not change the output values stored in the data files, in place of the full modules).

An important related advantage of modular systems is the ability to create and use reduced-form versions of the modules. Reduced-form versions approximate the more detailed modules in translating input variables (e.g., prices for demand or quantities for supply) to output variables (e.g., quantities and investments for demand and prices and investment for supply). Reduced-form versions can be easily substituted for the full modules. They would not be independent models, but simple mathematical structures estimated from the original modules to approximate these modules' full responses.

Such reduced-form versions would allow great flexibility in the use of the NEMS. When outputs from particular full modules of the system were not important, these components could be replaced by their reduced-form versions. If many such substitutions were made, the NEMS could be quickly run. Frequent fast runs would be particularly valuable when NEMS were being used for probabilistic analysis. In addition, when many reduced-form versions were used, users could routinely conduct in-depth analyses using the full module of interest without being burdened by the computational costs of running all the full modules. The committee believes that the NEMS could represent crucial relationships in the overall energy system using reduced-form modules, with little deviation in projections from those of the full modular system. Because these simplifications can introduce further error, work needs to be done to assure that the increased variability attributable to the use of reduced-form modules is held to acceptable levels.

Modular structure could allow the testing of uncertainty, including that associated with the structure of alternative modules. Simply removing one model and replacing it with another of a different structure would allow testing the uncertainty associated with alternative beliefs about various components of the energy system. Similarly, modularity would allow NEMS to integrate results from external models whenever this were desirable.

The committee below recommends that NEMS be developed to run on personal computers if the resulting configuration would not significantly sacrifice the NEMS models' content. Modular architecture could make this goal easier to achieve inasmuch as only some parts of the entire system would need to be in computer memory at a given time.

Disadvantages of Modular Systems

Modular architecture can create coding difficulties not seen in single integrated models that run with one kind of software. The modular structure proposed for the NEMS would rely instead on a control module to guide the operation of the other modules. Off-the-shelf software may not be available and EIA will probably need to develop such a control module.

Decentralized development of the NEMS could well lead to few (if any) people understanding the entire model. Thus, few may be capable of interpreting and assessing it as a whole. This is a legitimate concern. The committee believes that EIA should explicitly structure its efforts to avoid this difficulty. In particular, one or more EIA analysts should be charged with overall knowledge of all modules, even those not developed in EIA.

Time and effort should also be devoted to ensure that the model as a whole--or larger aspects of it--are understood by those with an appropriate range of strong expertise in modeling and energy-related matters, but not necessarily strong modeling backgrounds. It should also be noted that a modular design places a heavy burden on the architects of the system to think through and specify the requirements of individual modules, and to interact with module-builders in a continuing process of validation and modification.

Decentralized development could lead to significant quality control problems in both model development and model maintenance. EIA must take responsibility for maintaining high quality, even for modules developed outside of EIA and for making sure that EIA personnel fully understand them.

INTEGRATING MODEL OPERATION

The integrating system would take a set of inputs--for example, world oil price, U.S. population growth, and policy and other assumptions--and run the modules until the system converged to an equilibrium. Results from the integrating framework would then be used as inputs to run satellite modules.

The solution of the integrating model would typically represent market clearing, that is, supply being equal to demand, for every energy commodity, in every region, at each time. To find a solution, the modules would take as inputs price, quantity, and investment, or some other limited number of vectors from common data files. Once these modules were run, outputs--values for price, quantity, and investment, or other vectors--would be passed back to the common data files. These vectors would be indexed by energy commodity, region, and time. Environmental impact vectors could also be passed to and from the modules in finding a solution.

Figure 3-2 illustrates one convergence to an equilibrium, for energy price and energy quantity. The demand module response is represented by the downward sloping curve, which projects the energy quantity demanded at each energy price. The supply module response is represented by the upward sloping curve, which projects energy supply price for each energy quantity supplied. Equilibrium is represented by the values for price and quantity at which the two curves cross.

In Figure 3-2, an initial price is chosen as P_0 (for illustration, P_0 is chosen far away from the equilibrium price). The price P_0 would be read from the common data files by the demand module. The demand module would then project Q_0, a value for energy quantity that would be fed into the common data files.

The supply module would then read the quantity Q_0 from the files and provide a supply price of P'_0, which would be sent back to the common data files. The integrating system would be at a solution if P'_0 were close enough to P_0. If not, the next iteration would use both these prices to calculate a new price estimate, possibly a weighted average of the input and output prices. This new starting point is represented by P_1. The process is repeated, giving an output price of P'_1. And so the process would continue until input and output price values were sufficiently close.

Figure 3-2 illustrates an adjustment process such that energy prices and energy quantities converge to the equilibrium. Yet in other modeled adjustment processes, no equilibrium may be reached, but instead proposed solutions oscillate around the equilibrium point. For example, if the starting point of each iteration were set equal to the output price from the last iteration, the proposed solution could oscillate and would always oscillate for some shapes of the supply-and-demand curve. Additionally, when the number of prices and quantities is greater than those in Figure 3-2, the process of convergence to the equilibrium may be more complex. This complexity may slow the process of convergence or may prevent convergence from being reached.

To avoid such problems, EIA should adopt or develop appropriate algorithms for NEMS. For example, by choosing an appropriately weighted average of the initial prices (P_0) and the output prices from a given iteration (P'_0), convergence can be guaranteed for most supply and demand curves. The appropriate choice of weights also generally greatly decreases the time to convergence. EIA should devote some attention to adapting or developing algorithms to ensure efficient convergence to equilibrium.

In Figure 3-2 only a unique equilibrium, or single equilibrium, point exists. Because the supply module response is upward sloping and the demand module response is downward sloping, these two curves can cross only once: the equilibrium is unique. However, if there were many prices and quantities, there might well be more than one equilibrium.

If there are multiple equilibria, policy analysis and forecasting becomes especially difficult. With multiple equilibria, the model typically provides no indication which of the several possible system equilibria would be obtained. Therefore, unless the analyst can

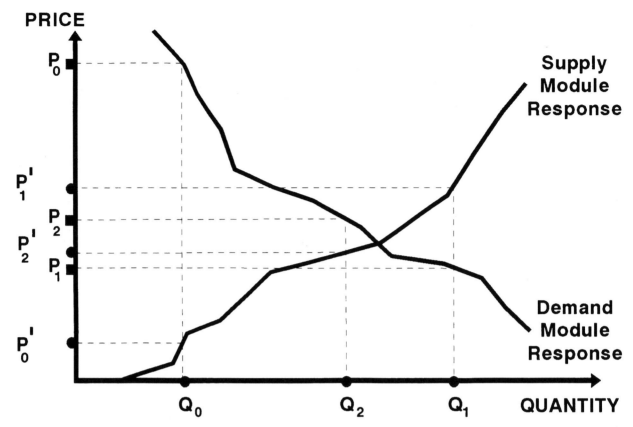

FIGURE 3-2 Illustration of the convergence of price to a supply-demand equilibrium.

assess which of the equilibria is most likely to occur, the projections would remain ambiguous.

In addition, policy analysis is hampered when multiple equilibria exist. Unless each equilibrium point is examined, the projected impacts of policy measures can be extremely misleading. Figure 3-3 abstractly illustrates the difficulty for the situation in which two outcomes, labeled "Variable 1" and "Variable 2" are being examined.

In Figure 3-3, for the first policy measures, the group of points labeled "equilibrium set" could all be equilibria, while for the second policy measures, the group of points labeled "equilibrium set B" could all be equilibria. Note that equilibrium set B is translated downwards and to the left from equilibrium set A. The interpretation is that, whatever

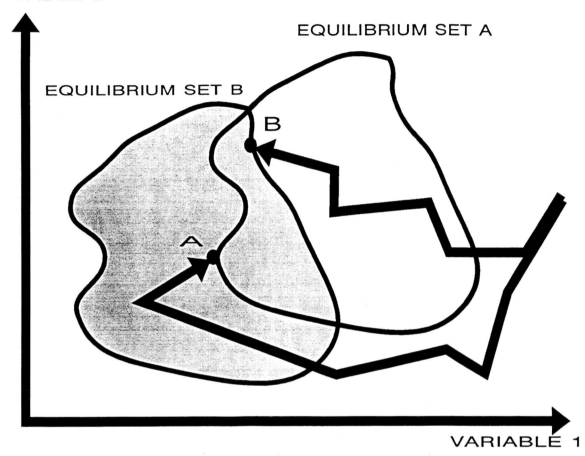

FIGURE 3-3 Illustration of multiple equilibria.

equilibrium point in fact occurs, changing to the second policies will decrease the values of both variables.

But a model run does not trace out all the equilibrium points. Rather, in a given run, variables are sequentially calculated until one equilibrium point is found. This process is represented by the crooked lines converging to points A and B. The points A and B would then typically be interpreted as the outcomes occurring with the first and the second policies, respectively. The difference between points A and B would typically be interpreted as the impacts of the policy choices.

Notice that point B lies upward and to the right of point A. Thus, if these two points were the only ones that had been determined using the model, it would be natural to conclude that changing to the second policy would increase the values of both variables.

This estimate of policy impacts would be incorrect, as indicated by the shifting of the equilibrium sets.

When multiple equilibria exist, therefore, modelers must trace out the set of equilibrium points and examine how the set of points change in response to policy measures. It does not suffice to find only one solution for each configuration of policies. We do not know whether the NEMS will have a unique equilibrium or multiple equilibria. However, EIA personnel must determine which is the case.

There are theoretical results that suggest the NEMS might have a unique equilibrium. These results depend upon mathematical properties of the matrices of supply-and-demand derivatives. However, we do not anticipate that all of the assumptions required to prove uniqueness will in fact be valid for the model. For that reason, EIA must explore the model, searching for multiple equilibria, until it is determined that multiple equilibria do exist or until confidence is gained in the idea that the equilibrium is unique.

In developing the model, careful attention should also be paid to the hardware to be used in running the model. The committee believes that NEMS should be configured to run on personal computers if such a configuration would not sacrifice the content of the model.[1] Modular architecture could make such a configuration easier to achieve. The control module could be designed so to load the modules sequentially into personal computer memory. The control module itself would always remain in memory, while other modules and data files would reside on the hard disk, with only one module at a time in random access memory (RAM). As long as each module were small to be run alone (in conjunction with the control module), then the entire system could be run on a personal computer. For personal computers with sufficient memory, the control module could keep several or all component modules in RAM, running them sequentially, and not expending time or frequent disk swaps.

If the NEMs model cannot be adequately structured for use with a personal computer, then it should be configured to operate on a workstation using current software that allows transportability among computers.

If the system could be run on personal computers, it could be made widely available, increasing the number of analysts who understood, used, and tested it. Configuring NEMS for workstations would reduce the breadth of its use outside DOE and EIA, but it would still be more accessible than if designed for use only on mainframe computers. The committee believes that such model transportability is important for allowing non-EIA users to understand, critique, and offer ideas for model improvement. For this reason, the committee recommends that NEMS be developed to run on widely available computers unless the resulting configuration would significantly sacrifice the NEMS models' content.

---

[1] For use within EIA, the entire NEMS should perhaps be available on a central file server networked with individual personal computers or workstations. Our recommendation does not preclude that possibility. Such a configuration would allow many analysts to access the same version of the model simultaneously and would allow each access to the most current version, even when the model were being continually revised.

PROPOSED MODULES

Figure 3-4 shows the general proposed structure of the NEMS in more detail. Modules shown in double frames indicate models that, to the committee's knowledge, do not exist now within EIA.

Separate regionally disaggregated modules would represent various energy-consuming sectors (residential, commercial, industrial, and transportation). These modules would provide as outputs the quantities of the various energy forms demanded for the input delivered prices of energy.

Separate modules would also represent various energy supplies (gas, oil, coal, and renewables). These modules would additionally represent U.S. regional transportation of energy, so that supply modules would provide as outputs delivered prices. Single-fuel conversion processes (e.g., refining of crude oil into products) would be represented within the corresponding energy supply model (e.g., the model for oil). These supply modules would take as inputs regional demand and provide as outputs regional prices over time by energy form.

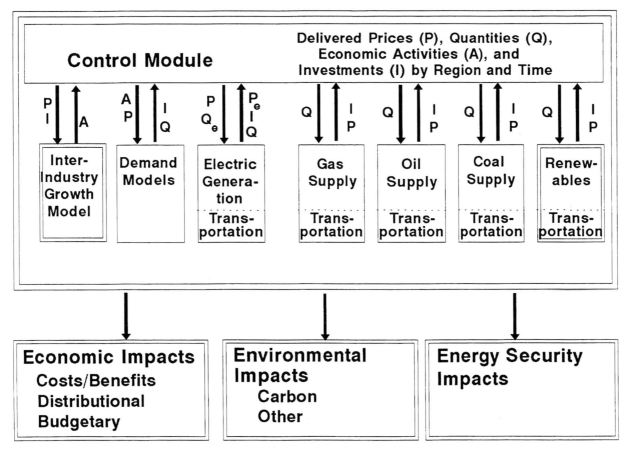

FIGURE 3-4 Proposed NEMS architecture in greater detail.

Electricity generation would be modeled taking as inputs the prices of the fuels used to generate electricity (oil, coal, nuclear, natural gas, and biomass) and the quantities of electricity demanded, and providing as outputs the prices of electricity and the quantities of the various fuels used.

A relatively simple aggregate model of the world energy system should also be developed, to estimate its interactions with the U.S. energy system.

The Control Module

The control module would include switches that would allow alternative versions of the modules to be used. It is envisioned that each module would routinely have three versions: the full model, a null model, and a reduced-form version of the full model. The user would choose which module version to use, and the control module would use this information in iterating to a solution. Such switches could be controlled with input data files or with more easily used menus.

Again, the control module would guide the sequential exercising of other modules. If the system could run on personal computers, the control module would guide the swapping of various modules into and out of RAM. It would choose the initial prices for each iteration based on the initial and output prices from the preceding iteration. It would also test for whether convergence had been obtained.

The control module would also include the user interface. The more "user friendly" this interface, the more readily the model could be transferred to users outside EIA and the more readily EIA personnel could learn to use the system. However, it is more important that the system have good content than that it have a user-friendly interface.

Fossil Fuel Supply Modules

The EIA now has fossil-fuel models for IFFS and other modeling systems. Additional work is ongoing. Because so much information is available about such modules, they are discussed here only with regard to a few points in medium-term forecasting.

The main concept behind current modules of energy supply is cost minimization, the obtaining of a given supply of a particular energy form at minimal cost. Analyzing and forecasting such activities often involves optimization models. For such models, delivered prices are represented by "shadow prices" or dual variables resulting from the constraint that fixed demands must be met.[2] Wellhead or mine-mouth prices are delivered prices minus the marginal transportation prices to the point of delivery.

---

[2]The terms "shadow price" and "dual variable" are synonymous. These concepts derive from optimization theory and can be interpreted as the change in the objective function (in this case, cost) that would occur if a constraint were relaxed (in this case, that a given quantity of energy were supplied). Also, see Kneese and Sweeney (1991) for papers related to non-renewable resources.

Such a representation implicitly assumes that the industry examined is reasonably competitive and that the information available is nearly perfect. It also assumes that replacement cost pricing is a prerequisite for new investment and that entrance and exit from the industry are relatively frictionless. However, if these assumptions are not valid, for example, if the industry does not minimize costs or if prices differ from marginal costs, then optimization approaches cannot be used directly without some adjustment. In particular, the optimization approach may ignore many important regulatory issues associated with the bulk transmission of electricity and transportation of natural gas. Particular attention should be devoted in NEMS modeling to such transportation issues. The approach to be used must be designed case by case.

Although oil and gas supply models have been used extensively for many years,[3] they may have substantial shortcomings (Huntington and Schuler, 1990; Bohi and Toman, 1984).[4] Current U.S. energy supply data bases and technology evolution models, particularly for oil and gas, are inadequate for all but relatively short-term analysis (not as great as 20 years). Oil and gas resources are presumed to be fixed and finite; technology is presumed to not respond to changing conditions. Overcoming these inherent problems in current long-term oil and gas supply modeling will require a two-part effort. First, the oil and gas resource base and supply models need to include the very large volumes of unconventional oil and gas that exist in low permeability formations, fractured reservoirs, coal seams and heavy oil and tar sand deposits. These resources already make a substantial contribution to energy supplies and will become even more important in the future. Second, the oil and gas models need to better capture the evolution of advances in oil and gas extraction technology. The general effect of these technology advances has been to improve the recovery efficiency of hydrocarbons and to decrease their extraction costs.

Similarly, the effects of improvements in oil, gas and coal extraction and conversion technologies are poorly captured by models. For example, continuing efficiency gains in coal mining, particularly through the now widely used long-wall mining technology, have steadily decreased the mine-mouth costs of coal and thus fuel prices for electric utilities below values forecast by EIA and other coal models. Also, technology advances in oil and gas extraction, including 3-D seismic exploration, horizontal drilling, well stimulation, sophisticated reservoir simulation and new drill bits, have all reduce costs and improved the efficiency of recovering new reserves. As a result, the supplies of oil and gas have been larger, as well as the costs to find and recover them lower than predicted.

While oil and gas are ultimately finite and depletable resources, when and how such depletion occurs thus becomes of critical interest to energy outlook and policy. The data base on long-term oil and gas resources appears to have major limitations (NRC, 1991b).

---

[3]In Figure 3-4 the oil and gas models have been shown as closely linked. Oil and supply models probably should be closely integrated because their economics are closely related. Oil and gas may both be produced from a single well or from different wells in the same field. Drilling equipment for the two is to a great extent interchangeable. Oil may be the target of drilling but gas may be found or vice versa.

[4]Current oil and gas models have even greater shortcomings for longer term energy projections, those beyond the year 2010.

In addition, the committee believes that changes in energy extraction technology and production costs will be important for understanding the potential of the very large unconventional oil and gas resources that are already making an impact on U.S. supplies.

Explicit modeling of such technology uses and implications and technologies' environmental effects will provide a stronger base for analyzing longer term energy supplies. The NEMS must recognize limitations in current oil and gas supply data bases for long-term energy studies and it should take steps to improve the oil and gas resource data that are essential inputs to such studies. Similarly, it will be important to differentiate policies that expand the available resource base from those that simply make it faster or less expensive to extract. In particular, considerations for NEMS should include adding the large in-place volumes of unconventional oil and gas resources into its resource assessments and supply models, and develop the capability to estimate the effects of continuing evolution in extraction technology and production costs.

The current EIA coal model is extremely detailed, far more so than would be appropriate for NEMS. One priority for NEMS development would be a great simplification of this model to use in general modeling, forecasting, and analysis. The simple model would then be used in NEMS. Detailed analyses of coal issues should probably be conducted outside the NEMS.

Energy Conversion Modules

The electricity generation model would be a hybrid of demand and supply models, taking as inputs the prices of possible fuels and the quantity of electricity demand (Q_e), and producing as outputs the price of electricity (P_e), investments, and quantities of fuels used to generate electricity. Other energy conversion activities involving more than one energy form in a substantial way (e.g., synthetic fuel conversion) would also be represented by hybrids of demand and supply models. However, such other modules will not be addressed here.

Output prices from the electricity generation model should not be represented as marginal costs of electricity generation and transmission. Most utilities are still regulated based on average cost plus a regulated rate of return on invested capital. For such sectors, a cost-minimizing approach could still be used but dual variables cannot be directly used for price determination. Rather, average costs should be calculated based on utility accounting conventions and used as estimates of delivered costs. Thus, the financial accounting information should remain in the electricity generation model.

The effects of utility-financed energy conservation investments should be considered when calculating average costs for electricity generation. Some such investments are simply expenses by the utilities while others are capitalized into the rate base and thus earn a return over time. A smaller number are paid for over time by the utility customer. Judgments must be made about the fraction of investments to be treated in each of the three ways. Investments either expensed or capitalized by the utility will increase the average price of electricity charged by the utility.

Modeling the electricity generation sector will present other special challenges for EIA. EIA should also give explicit attention to the changing structure of the industry. Demand-side investments by electric utilities and others should be addressed. Renewable sources and nonutility electricity generators who sell to utilities are growing in importance and present an additional modeling challenge. Financial transfers and utility incentives to end users should likewise be incorporated. Such structural changes in the sector imply that its behavior in the future may be very different from that observed in the past.

EIA's current electric utility model has a vast amount of detail about conventional fossil fuel inputs to electricity generation. EIA should simplify the representation of these inputs and otherwise streamline the existing model for the NEMS. For nuclear power, EIA currently estimates generation capacity and financial information offline, which serve as input to its current electric utility module. Additionally, the current model has little or no detail about renewable energy sources. In the following section, some approaches to modeling renewable energy conversion are addressed.

Renewable Energy Conversion Modules

Data requirements and acceptable model structures for renewable energy are different from those for fossil energy. For purposes of this discussion on how DOE models treat renewables, renewables are divided into two categories: biomass energy, for which ongoing fuel expenses are a significant fraction of total costs; and nonbiomass renewables like hydroelectric, solar, wind and geothermal energy, for which there is near total substitution of capital for ongoing fuel expenses. The latter type of renewables will be discussed first.

"Screening curves" are used in large-scale energy models to simulate electric utility choices of conventional primary energy sources: estimates of fixed costs are considered along with a measure of conversion efficiency and fuel prices to project end-use cost. For example, most electricity supply models specify a technology by capital cost in units of dollars per kilowatt ($/kW); fixed operating and maintenance (O&M) costs in units of dollars per kilowatt per year ($/kW-yr); variable O&M costs in cents per kilowatt-hour (¢/kWh); and efficiency or "heat rate" in units of British thermal units per kilowatt-hour (Btu/kWh). These factors are combined with fuel prices in units of dollars per British thermal unit ($/Btu) and a depreciation factor expressed as a "fixed charge rate" (the percentage of capital costs to be recovered each year). The output of this analysis is a screening curve with production costs in cents per kilowatt-hour (¢/kWh) on the Y-axis and "capacity factor," or percentage of design output over time (usually annually), on the X-axis. These screening curves and electricity demand functions are input to production cost models to simulate utility decisions about the use of existing resources ("dispatch") and thus fuel mix, and choice of new resources. Uncertainties in the screening curves and demand functions are treated parametrically over multiple model runs.

Conventional screening curves are inadequate to simulate choices of nonbiomass renewables, for three reasons, especially when these curves are aggregated over a generic resource base. First, because almost all production costs are fixed, a curve relating production costs to capacity factor conveys less information than the capital cost estimate itself. The shape of the screening curve makes the utility's acquisition choices appear to be very sensitive to demand function uncertainty. In reality, the opposite is true because the individual plant size for a renewable resource is usually small. Second, because there is no single market-clearing price for the primary resource (falling water, blowing wind, shining sun, etc.) and the resource spatial density distribution is highly variable, an average capital cost in dollars per kilowatt ($/kW) is almost useless as a production cost model input. For example, the output of a wind turbine is proportional to the square of the wind speed, and average wind speed varies by an order of magnitude or more over very short distances. Thus, production cost is very site-specific and highly dependent on the resource energy-density. In contrast, market-clearing prices smooth out production cost variance for natural gas and allow a combined-cycle gas turbine to be accurately modeled with a simple screening curve. Also, gas energy-density, transportation costs, and location-specific fixed costs are widely known and easily modeled. Third and finally, diurnal or seasonal variations in the renewable resource complicate the use of capacity factor as an independent variable.

Using traditional screening curves to model utility choices about the use of renewable resources induces several problems. Near-term utilization would tend to be underestimated because low-cost or high-value applications would not be captured in traditional screening curves that are based on average cost and value, even though they would be seen by the utility.

Such a modeling approach would also tend to overemphasize the importance of R&D for initial market penetration for the same reason. Barriers to commercialization would appear to be much larger than they really are. Such models would tend to show policies such as tax credits or capital cost subsidies to be less effective (subsidies would appear to lead to less incremental production than would actually be the case). Since many new technology penetration models imply that adoptions in one year positively influence adoptions in the next year, if early adoption were underestimated the error would be compounded over a long horizon.

Economies of scale could be misrepresented as increasing unit equipment size instead of increasing equipment production rate: cost reduction does not take place in the laboratory, but in the factory and the field. At the same time, such simulations would tend to overstate renewables' long-term contribution to energy supply because of inadequate representation of daily and annual cycles of energy production, low energy-density in the marginal resource base, and overly optimistic projections about R&D results.

These problems can be minimized by properly modeling capacity additions as a supply curve, recognizing low costs where resource energy-density is high and the value of niche markets where special characteristics confer high value. These high-value markets would support early high-cost production leading to lower costs as experience is gained.

Data requirements for renewables are therefore significantly different from those for fossil fuels. More attention must be paid to near-term markets and low-cost pockets of the primary resource. Energy production cannot be represented as a simple capacity factor for most renewables. More site-specific and technology-specific representations must be generated in a format that can be input into a production cost model. The model needs to be chronological rather than driven by summary representations of load shape. Because equipment unit production volumes for dispersed renewables like wind and solar energy are very high, learning curves and factory operations research methodology need to be considered rather than one-shot construction economies of scale in modeling cost reductions.

Biomass energy creates special modeling problems. Conversion facilities are like lignite plants in that both use solid feedstocks with low energy-density. Lower pollution control costs for biomass are balanced by this lower energy-density. Adequate representation of feedstock costs is the crucial problem. Low energy-density implies that collection and transportation costs are relatively high. Agricultural processing wastes and forest product residues represent one class of feedstock. Costs for these byproducts are very low and net energy balance is unimportant because the growing, collection, and transportation costs are absorbed by higher-valued products such as food, lumber and paper pulp. However, the biomass resource in such cases is limited.

Once the leap is made to growing biomass specifically for energy production, net energy balance becomes a key consideration. The costs associated with low energy-density drive biomass energy programs to high yields per unit land area. High yields drive the process toward high input agriculture. For these reasons, useful models and data are generally found outside the energy sector. Uncertainty within such models may be particularly high because the difference between energy input and energy output is the small difference between two large numbers.

In addition to the engineering and economic issues discussed above, there are significant policy issues that must be addressed. For example, the adoption of most renewables is shaped by widely varying state regulatory policies. The success of renewable energy development has depended on regulatory decisions to require utilities to purchase independently generated power for a fixed, relatively high price. In the future, such decisions could be made on an environmental basis in addition to an economic one.

The discussion above suggests that models to simulate renewable energy conversion will require a high degree of geographical and temporal detail. Such detail would lead to a model far more complex than is appropriate for direct integration with the NEMS. For this reason, EIA should develop smaller, simpler modules (but modules larger than reduced-form versions) that simulate the operations of the more complex renewable energy conversion modules. These smaller modules would be integrated with NEMS.

The difficulties of modeling renewable energy conversion suggest that the endeavor requires creative research, modeling, and data development. Significant intellectual resources, from EIA, other DOE offices, and outside DOE should be applied.

The International Energy Module

The price of oil in the United States is primarily determined by the world oil market, the availability of natural gas is influenced by Canadian natural gas exports, methanol prices in the United States are greatly influenced by world methanol prices, and the world price of oil and other energy forms is influenced by U.S. supply and demand: for these and similar reasons, a relatively simple aggregate model of the world energy system should also be developed, to estimate interactions between the U.S. and world energy systems (see, e.g. Manne et al., 1985).[5]

In particular, U.S. energy exports (e.g., coal) and imports (e.g., oil, Canadian and Mexican gas and electricity and methanol) should be included. It would also be useful to estimate energy supplies imported from and produced in the most unstable regions of the world. The goal would be to examine prices and quantities of U.S. energy imports and exports and to assess variables particularly important for energy security. The immediate goal in NEMS development would not be detailed examination of world energy markets but only the examination of their features most important to inform U.S. policy analysis.

Endogenous (determined internally by the model) determination of world energy prices and U.S. import and export quantities depends partly on reasonable representations of the impact of U.S. imports and exports on world prices. But if such representations exist, they could be incorporated in the integrating system of the NEMS model.

The international model of U.S. energy imports (oil, natural gas, methanol and electricity) would resemble other supply models, with world energy prices as outputs and U.S. demands for imports for the various energy forms as the inputs. The international model of U.S. energy exports (coal) would resemble other demand models, with export coal demands as outputs and the prices of exported coal as inputs.

Such representations would allow the assessment of policies expected to change the world price of oil or of other energy forms. For example, if increased U.S. oil imports led to increased world oil price, then from the U.S. perspective, the incremental cost to the United States of additional imports is greater than the price paid by the users of oil. This difference between incremental cost and price reflects the terms-of-trade costs to the United States of increased import prices associated with increases in the import quantity. There is disagreement among analysts about the degree to which terms-of-trade changes should be included in welfare assessments of policy actions. However, if one could adequately model the forces determining world oil prices, then such terms-of-trade impacts could be consistently quantified if the representation of world oil price determination were incorporated in the NEMS' integrating module.

It is more complex to model the influence of U.S. regulatory policies on the technologies that shape energy supply and demand abroad (for example, when U.S. standards for automobiles or appliances are applied internationally to these commodities). Yet if the impacts of such policies on world energy supply-demand patterns could be assessed, then such a representation could be used to estimate other similar policy impacts on world energy prices.

---

[5]This is easier said than done, for example, there is no accepted view about the structure of the international market, including whether the Organization of Petroleum Exporting Countries has ever exercised market power over the price of oil (Compare MacAvoy [1982] to Griffin [1985]).

A key difficulty in developing such an international module is understanding the nature of the relationships between U.S. actions and international conditions. This problem may be especially acute for the world oil market, because it is uncertain whether increases in oil imports raise world oil price, decrease it, or leave it virtually unchanged (Sweeney, 1981). The uncertainty stems from different perspectives about the workings of the Organization of Petroleum Exporting Countries (OPEC). For example, many economic models of OPEC suggest that the greater the demand for OPEC oil, the greater the increase in OPEC price. A significant decrease in demand would thus reduce price. However, OPEC might also be seen as an imperfect cartel that is more cohesive when oil prices are high. When oil prices are low, revenues to member countries are low and they may be less willing to reduce production. This model would imply that the higher the price, the lower the OPEC supply of oil, the reverse of the common assumption. For such reasons, it is difficult to have confidence in projected U.S. policy effects on world oil prices.

The modeling of natural gas, electricity, and coal imports and exports will probably be less difficult. The demand for U.S. coal exports can be expected to be negatively related to export price: higher prices decrease export demands. Imports of natural gas from Canada to the United States will be positively related to price: the higher the price, the greater the natural gas imports to the United States. Such relationships could be estimated for NEMS use based on more detailed models of the world coal market and Canadian natural gas market, even though the more detailed models themselves would not be incorporated in NEMS.

The Interindustry Growth Model

As far as the committee is aware, the EIA has not included interindustry, economic growth models in their modeling system, but has used macroeconomic models, such as the Data Resources, Inc. (DRI), quarterly macroeconomic model. Such quarterly macroeconomic models emphasize the aggregate demand for goods and services rather than the supply side of the economy. They are most appropriate, therefore, for relatively short time horizons. But for modeling over the roughly 10 to 25 period, analysis that emphasizes the ability of the economy to produce and considers the economy-wide structure of industry is more appropriate. Such models have already been developed and are broadly used (Jorgenson and Wilcoxen, 1990). For this reason, NEMS should probably acquire and incorporate an interindustry growth model rather than develop its own estimates of energy system interactions with the economy.

The interindustry economic growth model should incorporate the idea that economic activities both influence and are influenced by energy sector outcomes. For example, energy demand in the industrial sector varies significantly with the balance between heavy and light industry. Thus, the model should consider appropriate economic activities as driving variables for demand sectors. Economic growth rates will be influenced by the magnitude and type of capital investment required for energy supply development or in end-use technologies. In addition, energy prices will influence the relative sizes of economic sectors, the overall savings rate, and the patterns of capital investment. These relationships should be represented by the model.

Integrating an interindustry growth model into NEMS would require careful analysis of the linkages between that model and the rest of the system. For example, a typical interindustry growth model would include many of the energy industries that NEMS would be modeling in greater detail than would the interindustry model. An interindustry growth model such as the Jorgensen-Wilcoxen model would already have a full structure for determining energy demand. It is not clear that a full structure of econometrically estimated demand equations could be reconciled with bottom-up models of industrial energy use. At the minimum, major problems of logical consistency can be expected when the process of integration is started. In addition, the problems that will arise in matching the industry classification adopted in the interindustry growth model with the energy accounts within NEMS can be expected to be very difficult if not intractable.

Significant intellectual effort will be needed to correctly integrate an interindustry growth model into NEMS. And such an integration may ultimately not be possible at all and will not be possible within the next year. However, we believe that intensive modeling and research efforts could lead to a successful integration. EIA should structure a process that would bring the requisite intellectual talent to bear on this problem in the expectation that some integration can be achieved.

Energy Demand Modules

Two broad approaches are feasible in developing energy demand forecasting models: "top down" and "bottom up." Top-down models rely on aggregate relationships between energy consumption and economic, demographic, regulatory, and other forces (Hogan, 1989). They primarily use statistical techniques to estimate historical relationships between underlying forces and observed consumption (for a comparison of different models of this type see Bohi, 1981). Bottom-up models construct estimates by adding up the amounts of energy used in different energy-consuming activities (McMahan, 1986; EEA, 1982). They rely primarily on forecasts about future activities expected to use energy: the nature of such activities, their extent, the technologies available and likely to be adopted for these activities, and the behavioral rules that govern the evolution of these factors.

Current energy demand forecasting often combines the two approaches (e.g., Difiglio et al., 1990). For example, in estimating gasoline demand, an econometric model might be used to project vehicle miles traveled. Efficiency of cars can be projected based on technological feasibility and regulatory constraints, such as the corporate average fuel economy (CAFE) standards, as well as empirically based estimates of new car sales for each year. Estimates of vehicle miles traveled and fuel efficiency are combined to project gasoline demand for automobiles. The committee expects and encourages EIA to use such hybrid forecasting for much of its energy demand modeling.

Current EIA energy demand models are disaggregated by consuming sector rather than by fuel. Disaggregation by consumer sector is appropriate for several reasons. First, there is too much difference in energy use among the consumer sectors to aggregate them meaningfully in one model for many purposes. Second, interfuel substitution is an important factor: the price of one energy form can influence the demand for other energy forms within a single consuming sector. For this reason, it would be less useful for NEMS to build separate models for each energy form.

Current EIA demand forecasting models stress the response of energy consumers to energy prices and other economic and noneconomic driving variables. These models have been developed to be consistent with historically observed behavior. EIA energy demand models vary in their reliance on top-down estimates of observed energy consumption and bottom-up calculations of consumption. But all currently have very little technological detail.

Both bottom-up and top-down models begin with the observation that energy is typically not desired as an end in itself; rather, it is employed along with appliances and other machines to generate energy services (e.g., warm or cold air, lighting, and power applications). The demand for energy therefore depends on the stock of energy-using equipment, the energy efficiency of this equipment stock, and its rate of use.

One important implication is that, once energy-using equipment is in place, short-term variations in energy demand owe primarily to changing rates of use, whereas over the medium-term, both equipment stock and equipment energy-efficiency may change. Thus, when modeling energy demand, it is imperative that short-term dynamics be distinguished from medium-term dynamics. This distinction must be recognized in both bottom up and top down modeling.

Bottom-up demand modeling typically estimates energy consumption as the product of an activity times an energy- intensity, for each sector. The challenge to the modeler is to project the activities and energy-intensities. Both should be modeled as functions of energy prices, regulations, economic growth, and other forces. Bottom-up forecasting requires specific technology projections and projections about the underlying activities, such as construction of housing units, industrial production, and travel. Such analysis also requires analysis of the rates of technology adoption, conditional on the underlying forces.

Optimization as well as forecasting models can be used to analyze energy consumption. For example, some members of the committee believe that the energy-intensity of most equipment in use today, both new and existing equipment, is higher than "optimal" given the consumer's cost of energy and the cost of capital. Thus, one interesting kind of policy analysis might address whether private decision makers are using technologies in a manner inconsistent with their own self interests or whether the analysts have come to mistaken conclusions (perhaps by ignoring important subjective costs and benefits of energy-using activities or by an overly optimistic view of the technologies they are assessing).

If real-world market forces lead individuals to use too much energy, then models could be used to assess the economic, energy, and environmental effects of reducing the differences between actual and "optimal" levels of energy-intensity. To conduct such technology- and activity-specific analyses would require the characterization of a vast array of technologies. Many of these technologies would only become available at higher energy prices.

The bottom-up modeler needs extensive data, projections, and information:

o Data on energy use and projections of activities by sector, energy form, time, region, and other key characteristics for each possible energy price scenario.

o Data and projections about characteristics of energy-using technologies, such as their applicability, costs, performance, and environmental impacts, including operational variations and energy-efficiency add-ons, for both existing and prospective technologies.

o Data on, behavioral information about, and models of decision makers and decision making in the acquisition of energy technologies, including such factors as institutional setting, financing issues, response to reliability, and technical sophistication, as well as price. Such information should cover not only the current institutional setting, but also plausible future settings, including the availability of financial assistance and information to improve efficiency.

EIA is familiar with the conventional data requirements in these areas and is working to collect more current data, for example, for commercial buildings and especially industry, where the heterogeneity of energy-related activities makes good analysis complicated. Depending on the type of analysis, highly disaggregated data of the first two types above may be needed. Such data are receiving attention in other agencies and EIA can use this work, although the data surveys are limited, especially for the industrial sector. However, EIA would need to verify the quality of these data, since field measurements are inadequate for some of the data collection activities (e.g., no one has verified the accuracy with which census energy-consumption forms are filled out. Since there are many different types of units there may be a serious problem.).

Top-down, dynamic modeling of energy demand can itself be addressed in at least two ways. One body of literature, beginning with the classic study by Fisher and Kaysen (1962), explicitly introduces measures of the energy-using capacity of the capital stock, and then models factors affecting the use of this stock. A second approach, also considered by Fisher and Kaysen, has been further developed by D. McFadden (1983), Dubin (1985) and McFadden and Dubin (1984).

The discussion here focuses on aggregate sectoral models for areas in which it is impractical to obtain detailed data on a vast variety of energy-using capital goods, that is, on models incorporating the effects of the capital stock indirectly. Taylor et al. (1984) call these models "flow-adjustment models," since changes in the flow of energy depend on the speed with which capital stocks are adjusted through investment.

According to flow-adjustment models, consumers form expectations concerning future prices and incomes, and based on these expectations and other exogenous (external) variables, they choose desired or preferred levels of energy consumption. At any point in time, consumers' actual purchases of energy may differ from their desired consumption, in part because of psychological, institutional, and technological factors that make instantaneous adjustment costly.

To use a flow-adjustment approach in a top-down model, therefore, requires that several issues be addressed:

o The factors affecting desired levels of energy consumption must be specified explicitly, including the way in which consumers' expectations of future prices and incomes are formed.

o The adjustment process from actual to desired consumption must also be formulated.

The second issue is typically approached through the partial adjustment model, in which current energy consumption depends on past energy consumption and other exogenous variables.

As for the first, desired energy consumption can plausibly be argued to depend on expected real energy prices and expected real income (or economic output), particularly because energy is used in conjunction with long-lived durables, demand for which in turn depends on expectations of several explanatory variables. Since expectations of these variables are usually unobserved, the challenge is to specify a working representation incorporating nonstatic expectations, and in terms of observed variables. This is the "adaptive expectations" approach discussed in most econometrics textbooks, but it can be expanded to include optimal or rational expectation predictors of real energy price and real income or output (see Nelson and Peck [1985] for an example of such estimation, and Pindyck and Rottenberg [1983]).

The most modern demand models constructed by Daniel McFadden and his research associates employ engineering-economic models, sophisticated econometric and statistical procedures, and extensive data bases on individual U.S. households for the residential demand for electricity (Goett and McFadden, 1984; Berndt, 1991; the following discussion derives from Berndt). These models are based on the conceptual framework of Houthakker (1951) and Fisher and Kaysen (1962). This engineering-econometric model employs data from the 1977-79 U.S. Department of Housing and Urban Development Housing Surveys, an annual survey based on over 70,000 households. These surveys include information on household's appliance stocks and sufficient geographic information to construct normal weather condition variables and the marginal price schedules facing these households. Other microdata from the U.S. Department of Labor Consumer Expenditure Survey, as well as engineering and construction sources, were also used in this collection of models referred to as the Residential End-Use energy Planning System (REEPS). Such use of microdata can allow a detailed understanding of the variables affecting appliance choice that cannot be discerned using aggregate or average data.

One module in REEPS represents appliance choice decisions based on household socioeconomic characteristics, appliance holdings and attributes, type and size of residence, and geographic and economic characteristics associated with the household location. This appliance choice framework was modeled separately for each energy-related function such as space heating, water heating, or refrigeration. The model captures weighing of capital and operating costs in making household appliance decisions. The econometric technique involves a discrete choice framework known as the multinomial logit model and its generalization.

The second major modeling effort in REEPS involves households' choice of utilization rates with specified short-run, nonlinear demand equations that include explanatory variables such as appliance operating efficiencies, marginal energy prices, and demographic and geographic variables; equation estimation was accomplished using nonlinear least squares procedures.

Simulation of the REEPS models were conducted under a wide variety of alternative assumptions with the results used to summarize the sensitivity in the short and long run to changes in economic, regulatory and demographic variables. Elasticity estimates are derived from the simulations.

Behavioral Information Requirements

Both top-down and bottom-up models include behavioral information. However, bottom-up models require specific formulations of behavioral rules, while top-down models implicitly make behavioral assumptions through aggregate estimated relationships. The explicit assumptions of bottom-up models allow easy interpretation of the model results, but may not correspond to actual behavior observed. The implicit assumptions of top-down models may correspond in the aggregate to consumption behavior, but may prove misleading in forecasting the future, for example, if behavior changes, because of policy changes not reflected in price. In addition, when prices are the only relevant policy variables incorporated in such models, it is difficult to use these models to examine policy options that do not involve (implicit or explicit) price changes.

One kind of behavioral information concerns investment choices among types of equipment with different energy-efficiencies. Unfortunately, there are major gaps in our knowledge about such decisions. Some analysts assert that implicit annual real discount rates of 30 percent, and higher, characterize markets for energy-efficient equipment (Ruderman et al., 1987; Hausman, 1979). They believe that consumers give energy-efficiency much lower priority than simple economic analysis would suggest, and that producers and sellers often do not market products with energy-efficiency benefits. In a bottom-up forecasting model explicit assumptions must be made about whether these assertions are correct. If such assumptions are made, the modeler should understand quantitatively and in detail the forces that lead to the assumed implicit discount rates. With top-down models this issue is not addressed explicitly.

For example, a fundamental concept in bottom-up demand analysis is "energy service," a particular "good" in which energy consumption plays an important role. If the passenger transportation energy service is assumed to be mobility, then passenger-miles for different transportation modes constitute the activity and energy demand is the sum of energy demands for each mode. Unfortunately, the issue of what constitutes the fundamental service is difficult yet crucial for longer range modeling. People may not simply substitute one mode for another, for example, a mile of travel on a bus for a mile of travel by car. The fundamental service in daily travel may actually be access to various amenities. The suitability of different travel modes and trip lengths depends on the nature of the community in which one lives and works. The need for travel also depends on alternative forms of access, such as electronic access to information. A useful prediction of demand over several decades should thus raise fundamental issues associated with the amount of daily travel. Daily travel must be related to such factors as land use practices, public transportation investments, population densities in metropolitan areas, and alternative technologies for meeting access requirements. It is not realistic for all such information to be captured in NEMS and the committee is not recommending such an approach. However, a model that fails to address the most important of these issues will not answer crucial questions about transportation energy policy.

What are needed are the identification of key variables, insight about their behavior, and some lively debate about the ways modelers may improve their assumptions. EIA does not have to conduct or sponsor all the research needed. But it is essential that EIA allocate sufficient human resources to determine the principal behavioral information needs for effective medium-term demand modeling, and to determine what relevant information exists and what kinds of information require new surveys or, more generally, research. Where behavioral information is inadequate, EIA should help generate interest in the issue and follow through by soliciting research or stimulating another agency to sponsor research. Where funding is needed to spur essential research, EIA should have the resources to participate.

Sectoral Information Needs

Residential and Commercial Sectors. Top-down econometric studies have shown that residential demand for energy depends on a combination of economic, demographic, technological, weather, and seasonal factors.

Detailed technology-characterization of the residential sector has already been carried at Lawrence Berkeley Laboratory and the Electric Power Research Institute. Most energy services can be defined straightforwardly, and data can be collected on alternative ways to satisfy the energy service at different levels of technology and investment. Each technology, both commercially available and new technologies, have associated costs and energy savings. (The cost and performance of not-yet-available technologies are, of course, speculative.)

Using such analysis, a supply-curve (which may be quite uncertain) for space heating is constructed. The supply curve describes the rate of energy consumption as it depends on the level of investment in space-heating efficiency equipment. The level of investment can be related in turn to energy prices and decision-making behavior. Such supply curves have been incorporated into bottom-up models.

A critical need for the bottom-up approach in analysis of the residential sector is in modeling of the aggregate behavior of contractors who build residences. Information on the actual implicit discount rates characterizing the householder market is also needed.

Energy demand of the commercial sector has traditionally proved more difficult to model econometrically than that of the residential sector. One reason for the difficulty is that the commercial sector is extremely diverse, ranging from hospitals to retail stores to churches and offices. In addition, decision makers for investments in energy-using equipment are a highly heterogeneous group, ranging from speculative builders, who will not own or operate the buildings they build, to builders who will be owner-occupiers. This heterogeneity should be collapsed in the form of one aggregate relationship whichever modeling approach is used.

Industrial Sector. Industrial sector energy demand is greatly affected by the level of regional economic activity. Although the ratio of energy use to gross national product is important, it is not fixed, and it is now widely known that this important ratio varies with changes in real energy prices and other factors. Such changes, however, tend to take time and are often larger over the long than short run, as in the residential sector.

Energy demand in the industrial sector differs from that in residential and commercial sectors in a number of ways (Halvorson, 1978). In the industrial sector fuel substitutions, among electricity, gas, oil, and biomass, are more readily made. Substitution possibilities vary considerably among the various industries. In a top-down model it is thus important to estimate empirically the interfuel substitutions by including all energy prices as regressors in the industrial energy sales equations. Like the commercial sector, the industrial sector is very diverse, with different industries having sharply different energy consumption patterns. Aggregate industrial sector models are essentially confined to employing an average energy price; but marginal energy price for the "representative" industrial customer is a concept that does not make much sense.

For bottom-up modeling, the industrial sector presents a strong contrast to the residential sector. While rather extensive energy and activity time-series data are available for the manufacturing sector, these data are extremely limited given the heterogeneity of the industrial sector. It is not feasible to model industrial energy use reasonably in terms of generic energy services as it is in the case of the residential sector. A good analysis requires examining the different production processes and the potential for changing them, in energy and material terms. To model all production processes of interest would be too costly.

However, not only does considerable process detail still need to be described, technological and economic futures for each subsector need to be characterized. One approach to the challenge may be to model a small number of subsectors in some detail, selecting them for their importance and representativeness. For longer term modeling, the selected subsectors should not be confined to energy-intensive industries, but include industries that are likely to experience major growth. The analysis of even a few subsectors would require major data gathering and analysis including engineering case studies in the field for each subsector to establish a baseline for analysis, and wide-ranging efforts to identify and characterize potential process changes. Moreover, this information should be informed by a sophisticated appreciation of the likely economic future of any subsector, the evolution of its products and competitive climate (Boyd et al., 1986; Squitier, 1984; Doblin, 1984; Ross, 1984).

Table 3-1 indicates the nature of available data and information on the industrial sector. Some energy consumption and activity data are regularly developed by EIA and

TABLE 3-1 Available Data and Information, Industrial Energy Demand

Category	Data/Information Availability
Energy consumption	
Electricity	Four-digit manufacturing, annual
Fuels	Two-digit manufacturing, three-year; Some four-digit manufacturing, three-year
Production activity	At the four-digit level, manufacturing output available in nominal dollars annually; deflated output available after further delay; physical measures (tons) of production of selected materials available on a timely basis at least annually
Technology	
Characteristics	Extensive anecdotal but little survey
Basic process	information on technology type and
Existing	technology type and performance. few data on extent of technologies and trends in use;
Prospective	Little information
Conservation add-ons	Extensive information on technology types, but not on costs and adoption.

in the Department of Commerce's Annual Survey of Manufactures at the four-digit Standard Industrial Classification (SIC) level of manufacturing, except that fuel consumption is not sampled extensively enough to disaggregate well by sector. If adequate data were collected for disaggregation at the four-digit level, this would be satisfactory in most cases. There is potentially valuable data by four-digit SIC section and by geographic location from the National Acid Precipitation Analysis Program (NAPAP) and from Dun & Bradstreet data bases for the mid-1980s. Considerable effort would be needed to bring these data into useful form, however. Energy consumption data for nonmanufacturing areas (agriculture, mining, and construction) are very poor. Moreover, while there is much information on the claimed performance of major industrial process technology, there is little survey data on actual performance. Cost information on potential energy-efficiency improvements and on the extent of adoption of efficiency technologies are largely lacking in a form that could be used by modelers. The process of adopting energy-conservation technology is poorly understood, although there is anecdotal information on the role of high implicit discount rates in this process.

Analysts disagree substantially about even the most basic industrial sector modeling issues, including about how much price trends determine the creation and introduction of new technology, and about the proper way to measure production activity. EIA needs to develop a new industrial model and the modular structure proposed for NEMS proposed will allow EIA to change industrial sector models as they evolve and improve.

Transportation Sector. The transportation sector presents perhaps the greatest challenges of any sector for mid- and long-term modeling. Again, the type of community in which people live has a major impact on the extent and types of travel. Fundamental transportation fuel choices also face us in the coming decades. Petroleum supply, congestion, and emission crises may lead to relatively rapid and profound changes in policy and behavior and in turn to major unexpected changes in transportation technology and activity. Such dramatic changes also confront modelers in some areas of energy supply. But for the transportation sector, little relevant data and information are available and modeling approaches largely remain undeveloped. Present models are, naturally, based on past activities and technologies and past trends.

A good start has been made in analyzing past changes in the use of transportation fuels (Sperling, 1988) and the costs of a few new automotive fuel choices (U.S. DOE, 1988; NRC, 1990). The impact of gasoline prices on vehicle miles traveled has been estimated by many researchers. Much less is known about the effects on daily travel of type of community and land use public-transportation, and transportation management policies (e.g., parking charges, high occupancy vehicle lanes, and road pricing). Even in traditional areas of data gathering, knowledge is far from adequate. The sudden increase in vehicle miles traveled per driver in the late 1980s is not understood. It could be due to metropolitan area growth, changing employment patterns (with more outsourcing of services and part-time work), or increased income in high-income brackets. The effects of income itself are poorly understood, often being confounded with household size and employment effects.

There are data and information needs in all these areas. Current acquisition of data, for example, in the Department of Transportation's Nationwide Personal Transportation Studies, is slow and somewhat limited in its usefulness. The EIA's Residential Energy Consumption Survey (RECS) data series <u>Consumption Patterns of the Household Vehicles</u> is even more limited in its usefulness to the modeler, now that it no longer provides a reliable estimate of in-use vehicle fuel economy. EIA needs to develop more detailed transportation models that can better respond to modeling policy options.

Utility Demand-Side Programs. Promising opportunities for energy conservation are seen in policies that encourage investment by regulated utilities to increase customers' energy-efficiency. These policies bring utilities' financial criteria and expertise to bear on demand-side as well as supply-side investment. Several states, representing roughly 20 percent of the U.S. population, have already committed themselves to such policies through their utility commissions.

The effort has brought into focus the lack of information available to analysts about demand as compared to supply (Hirst, 1989). Electricity supply data are collected in great detail and with great frequency. EIA has annual electric utility reports relating to many aspects of generation, but none specifically on the recently developed demand-side programs. EIA publishes three surveys with highly limited information on end use, one for each major demand sector, every three years. However, the infrequent survey schedule and long delays in publication are less an impediment to demand analysis than is the paucity of detail.

It is likely that the energy use reductions created by demand-side programs can account for a substantial share of new electrical capacity requirements. For example, California has adopted as a goal that 75 percent of its projected energy growth be satisfied by demand-side programs. Analysis of this potential requires substantial data. A variety of data are beginning to be collected at the state level. Improved definitions and standardization of information collected and national collection of selected information are all needed (Hirst, 1990).

Information Needs for Policy Analysis

In addition to ongoing data and information, demand analysis requires specific information on the efficacy and other characteristics of selected public policies. For example, the effectiveness of different policies in encouraging investment in more efficient energy-using equipment needs to be understood. Specific policies include investment tax credits and regulatory mechanisms to reward utilities for investing on the consumer's side of the meter. The way such policies work is complex. For instance, detailed field studies show that the effects of the energy conservation tax credit in the early 1980s were much more limited than might be assumed in a typical model of investment decision making (Alliance to Save Energy, 1983). Similarly, utility demand-side programs need to be understood through research before such policies can be adequately modeled. Such programs vary from state to state and utility to utility.

The evaluation of federal policies appears to be an awkward research area for federal agencies to support, but means should be sought to foster this activity. Evaluations of the many energy-demand policies tried in the late 1970s and early 1980s, as well as those undertaken recently, would be especially valuable.

Quality and Timeliness of Data and Information

The value of data is obviously limited if they are not accurate or not available when really needed. On the other hand, because increased precision or speed of data collection entails increased costs, accuracy and timeliness must be understood relative to the eventual data use. Most statistical agencies specify the nature of data quality by providing details on how data are collected, measures of sampling variability, questionnaires used, validation methods (if any), and characteristics of sources.

Because energy issues and problems are sometimes urgent, preliminary estimates (with measures of uncertainties) that are available quickly can sometimes have greater value than more precise data available several years later. Such timely energy data are largely lacking for most areas of demand, except at the most aggregate level.

Data are available from organizations outside EIA, including industry sources, such as the American Petroleum Institute and American Gas Association, state and local government sources, such as the Northwest Regional Power Planning Council and California Energy Commission, and others. Use of data from such sources could greatly improve EIA analysis.

Reduced-Form Modules

As discussed above, EIA should routinely estimate reduced-form versions for each of the major components of the integrating system. Reduced-form versions would provide estimated response surfaces of the full modules. Relatively simple functional forms could be chosen to approximate the more complete models in translating from the input variables (e.g., prices for demand and quantities for supply) to the output variables (e.g., quantities and investments for demand and prices and investments for supply). A reasonably accurate approximation of this translation from inputs to outputs is necessary to capture the essential information captured by the complete model.

However, it is neither necessary nor possible that reduced-form versions exactly represent the full modules. How much variation between the two is acceptable depends on the modeling needs. For example, a reduced-form energy consumption module for a sector in a given region might include a base-case projection of energy consumption along with a constant elasticity (own-elasticity and cross-elasticity) approximation for the impacts of energy prices on consumption of different energy forms over time.

Reduced-form versions would be estimated for each region in the integrating system. Regional estimates would not be collapsed into one national estimate. Such a collapsing could be expected to significantly modify the nature of the feedbacks in the overall system in response to changes in one component of the system. Thus, collapsing several regions into one geographically aggregated reduced-form version could be expected to lead to significantly different projections than those resulting when a regional structure is used. However, the magnitude of the error associated with such regional aggregation cannot be estimated at this time.

Reduced-form versions should be periodically recalibrated (refitted) to pseudodata generated by running the full model. Recalibration would consist of fitting the parameters of the reduced-form version based on a structured set of runs from the full modules. The process is essentially a statistical procedure and statistical concepts can be used to guide it. However, to ensure that such recalibration can be carried out routinely, EIA must develop the software to facilitate the process. Such software probably does not exist, although research in this area has been conducted (Griffin, 1971, 1978, 1982). In addition, the statistical approaches to such reduced-form representations must be carefully understood.

In NEMS operation, reduced-form versions would typically be used for many or all sectors except those being analyzed. The full module would typically be used for the sector analyzed. Reduced-form versions would allow close approximations to the operation of the entire energy system with significantly decreased computational time and costs.

Satellite Modules

NEMS Satellite modules would examine the impacts of the energy system on economic, environmental, and security concerns. Many economic impacts would be evident directly from the outputs of the interindustry growth model. For example, overall rates of growth in gross national product (GNP) and the distribution of that growth among industries would be projected by the interindustry growth model. Other economic estimates, such as the income and regional impacts of energy policies, short-run macroeconomic effects, and effects on federal financial variables, such as the federal deficit budget, would be estimated through the use of satellite models.

Energy Security

Energy security problems are based upon the adjustment difficulty facing the U.S. and other economies in times of rapid changes, particularly rises, in the world oil price (Hickman et al., 1987; Toman, 1991). Rapid increases in oil price lead to unemployment (of people and capital equipment), GNP losses, inflation, and sharp jumps in costs of imports. Energy security, then, is measured by the likelihood and severity of economic losses associated with energy disruptions.

The severity of economic losses during energy disruptions is in proportion to the increase in the average U.S. energy price, which in turn is influenced by the fraction of the world oil (or energy) market that is disrupted. Therefore, security estimates will depend upon projections of oil production or production of close oil substitutes within unstable world regions, relative to worldwide production.

The severity of economic losses during energy disruptions also increases with the economic value of the energy (as a fraction of GNP) whose price suddenly increases. Thus the modeling system should be able to represent the dollar quantities of the various energy forms used in the United States as fractions of GNP. Short run economic analysis, most likely outside of the NEMS, must be performed to examine the linkage between world oil price changes during disruptions and changes in the prices of the various energy forms. Short run macroeconomic analysis, also probably outside of NEMS, must be conducted to examine the linakages between energy price changes and economic consequences.

The Environmental Module

The environmental implications of energy policy are of major and growing social and political concern, and they require significant attention in designing the NEMS. The integrating systems currently employed by DOE and EIA for national energy analyses do not contain an environmental component. However, some of the satellite models used by DOE for specific energy sectors (e.g., electric utilities) do reflect environmental factors and could serve as building blocks for a more systematic treatment of environmental issues.

Some of the basic requirements for a NEMS environmental module are sketched below, along with a strategy for developing NEMS capability in environmental analysis.

There is no single answer to the question of what environmental issues the NEMS should address. However, the committee believes that NEMS cannot be used appropriately to model highly site-specific environmental impacts nor those associated with unusual accidents. The system would be more appropriately used to estimate chronic environmental impacts associated with the energy system. For example, the model could be useful for estimating emissions of greenhouse gases, acid rain precursors, and the various ingredients of air pollution (volatile organic compounds, nitrogen oxides, and air toxics) outlined earlier in Table 2-1.

Such emissions are of concern because of their potential effects on natural ecosystems and human health. In most cases, however, robust measures of these effects are not readily available. Indeed, there are still significant uncertainties in most estimates of the environmental effects of energy production and use. The issues that the environmental model could address would not be comprehensive. Other environmental concerns--many in the context of local or regional issues, such as wastewater discharges, land-use issues, site-specific impacts--also are associated with energy projects. NEMS should be structured to examine the environmental issues that are expected to have a major influence in the development of U.S. energy policy in the next decade and beyond.

How, then, should the NEMS incorporate environmental concerns? The committee believes that the NEMS component models should have sufficient technological detail to estimate the total emissions of primary pollutants associated with component areas. A subsequent goal would be to develop auxiliary measures that relate emissions to effects at the global, national, and (where appropriate) regional levels. The last effort would involve experts and models largely outside EIA to characterize environmental effects and associated uncertainties.

The committee believes that all such environmental analyses should ultimately help provide the empirical basis for pursuing a policy goal of full social cost pricing, in which all the environmental "externalities" of energy production and use are valued in some appropriate way (whether using monetary or nonmonetary measures). Recognizing the difficulty of empirical work toward supporting this goal, however, the committee recommends an initial effort focused on quantifying the direct and indirect emissions associated with various environmental, health, and safety issues. This analysis would still allow interested individuals and organizations some insight into the environmental consequences of energy system changes (Edmonds, 1991; Edmonds and Barnes, 1990; Hogan and Jorgenson, 1991; Manne et al., 1990).

The approach envisioned would begin by characterizing the direct air emissions of primary pollutants for major energy supply, transport, conversion, and end-use activities. For example, emissions of carbon dioxide, sulfur dioxide, oxides of nitrogen, and trace species from coal-fired power plants, petroleum refineries, and other energy conversion sources could be characterized in terms of coefficients applicable to energy quantities calculated by NEMS for a particular scenario. This approach would yield an estimate of aggregate national emissions. Similarly, emissions from automobile use, residential heating, and other end-use activities can be estimated. This approach was used in some energy modeling in the early 1970s and has been updated in more recent linear programming models used in the United States and Europe. Appropriate data should be available through other organizations such as the Environmental Protection Agency (EPA); EIA need not develop the primary data itself.

A related, but more ambitious, effort would employ data available for major industrial processes to estimate the secondary or indirect emissions associated with energy-related activities. For example, in comparing solar photovoltaics to fossil fuels for power generation, one might want a more complete picture of environmental implications, including emissions from the mining and refining of materials for solar cells and other plant components and similar information about fossil fuel plants. These broader analyses would require something akin to an environmental input-output model of the U.S. economy--something not currently available at DOE or EIA. Yet sufficient information is available to begin constructing this type of model for use in conjunction with future energy analyses.

Inventories of direct and indirect emissions provide a starting point for quantitatively assessing the environmental implications of alternative energy scenarios (Reister, 1984). In the longer run, additional measures and models will be needed to assess the effects of different scenarios of projected emissions. In this area, again, the committee envisions NEMS drawing on other work in the public and private sectors to find answers, such as cooperation with the Environmental Protection Agency. For the short term, estimates of the national emissions implications of energy scenarios will add significantly to the analytical capability of NEMS.

It may well be that EIA cannot devote sufficient resources to the ambitious tasks described above. In that case, it is recommended that EIA initially focus attention on the greenhouse gas emissions associated with the extraction, transportation, and use of energy. This area will probably continue to command attention in the years to come. Experience in this area may then provide EIA with sufficient expertise to move to the more complete analysis outlined above.

Report Writers

The NEMS system should include well-designed graphics and report writers to provide routine graphical and numerical outputs of the types normally found to be helpful for analytical purposes. The outputs would take the form of standard projections of variables associated with energy security, the economy, and the environment. The system should also allow custom graphics to be generated easily. In addition, report writers should make runs self-documenting, by providing information such as the date of run, the settings for the various switches, and the model user.

NEMS COMPARED TO CURRENT DOE MODELING

Since the mid-1970s, several large-scale integrated modeling systems of energy supply and demand have been developed at EIA and later in the Office of Policy, Planning and Analysis, for use in recurring publications of energy forecasts and for policy analysis. The modeling systems include the Project Independence Evaluation System (PIES), later renamed the Midterm Energy Forecasting System (MEFS), the Intermediate Future Forecasting System (IFFS), the Long-Term Energy Analysis Program (LEAP), the Short-Term Integrated Forecasting System (STIFS), and Fossil2 (see Appendix E).

One trend exhibited in these models has been for increased modularity and decomposition. Another related trend has been the increased diversity of modeling methodologies employed in the modeling systems. These trends have evolved as system coverage and the complexity of energy issues treated have both increased.

The model structure proposed in this report resembles current EIA modeling systems, particularly the IFFS, which is a modular system. The proposed NEMS would represent a logical continuation of EIA model development. NEMS would draw heavily on current EIA approaches and could use revised versions of many current EIA models.

There would be some notable differences, however, between NEMS and the IFFS. Reduced-form versions of full modules have not been constructed. The IFFS system cannot run as a complete system on a personal computer, although parts of it can. NEMS would also require some entirely new models, in particular, the interindustry economic growth model and an environmental impacts model. In addition, current models of economic and energy security impacts are not adequate for the proposed system. Finally, IFFS calculations currently clears markets one year at a time, thereby precluding any representation of rational expectations or other "look-ahead" behavior.

The Fossil2 model of the Office of Policy, Planning and Analysis, which was used as the integrating framework for the first NES, is constructed using a different strategy. It is not modular. Rather, it includes all supply, demand, and market-clearing equations in one comprehensive dynamic simulation model. Each sector is represented by a set of differential and integral equations that relate the stocks (energy production, transformation, and consumption facilities and entities) and the flows (energy quantities, prices, and information). It would be very difficult, if not impossible, to substitute alternative component modules in Fossil2.

For the NES, the various component parts of Fossil2 were calibrated to mimic the EIA Annual Energy Outlook (AEO) reference case to the year 2010. Fossil2 parameters describing sectoral behavior were changed so that projections (at baseline prices) corresponded to the projections in the AEO base case. However, the response of supplies and demands to prices in general was not calibrated to the external models, such as those at EIA.

This use of Fossil2 would be quite similar to the proposed NEMS with all (only) reduced-form versions, if Fossil2 were calibrated so that the responses of outputs to inputs (for example, quantities to prices) were chosen to correspond to the external models. The key difference, however, is that Fossil2 components may well be more complex than NEMS reduced-form versions. Such greater complexity would provide greater richness of detail and more policy "handles" for analysts, but would make calibration more difficult for the NEMS. Nevertheless, NEMS reduced-form versions could be exactly like the components of Fossil2, if Fossil2 components were deemed the best reduced-form versions of the external models. Fossil2 components then could be used in NEMS if it were possible to separate them in the form of component modules.

The Fossil2 model differs from the proposed NEMS in that its structure is explicitly calculated on an annual basis: one year is calculated, then the next, and so on. Given such a structure it is easy to incorporate myopic expectations and adaptive expectations. However, to incorporate rational expectations, the entire model must be run over and over, using the output prices from any run as input prices for the next run. While in principle such an approach is feasible, in practice it would be very time consuming and probably not routinely pursued.

The Fossil2 model also differs from the proposed NEMS in its market-clearing mechanism. Prices move over time in response to differences between supply and demand, so that in any given year prices might not cause markets to clear. To minimize the modeling difficulties such nonclearing might cause, Fossil2 is recalculated four times for each year. While four interactions is probably sufficient to bring each year's supply and demand into close approximation, it cannot be guaranteed that markets would clear.

NEMS TREATMENT OF CONCEPTUAL ISSUES

NEMS architecture should incorporate conceptual issues that are significant to debates about energy market operations. This section considers how the proposed NEMS could address such conceptual issues.

Market Disequilibrium

NEMS architecture should be based on the idea that energy markets will clear: energy prices will take values such that supply is equal to demand for each energy form (Weyant, 1985). One question is the time it takes markets to clear, given whatever regulatory or other rules exist. The concept of equilibrium is less useful if markets do not clear within the time period modeled, for example, one year.

The equilibrium price model assumes that instantaneous adjustments occur; it does not consider the actual time it may take real markets to clear. In some cases, market clearing occurs over a long period as when there are government-imposed price controls and nonmarket allocations of energy forms. Government interventions can create market disequilibria that last for years, if not decades.

The NEMS modeling system could be used in a simple way to examine non-market-clearing associated with price controls. Rather than require the price-controlled market to converge to equilibrium, the model user could set product price--say retail price for gasoline--to the government controlled price. The test for convergence could be modified to allow the price-controlled product market to remain out of equilibrium while other markets converged to supply-demand equilibria (equilibria likely to differ in many markets from those that would obtain without the price controls).

Another important question is the market-clearing mechanism. Do markets clear based on fluctuations in price? With other clearing mechanisms (fluctuations in production, for example) the economy still has competitive markets that clear, but these markets have different characteristics and outcomes. Instead of moving to the equilibrium point given by conventional supply-and-demand curves, the markets may clear by shifts in the curves themselves. For example, when there are government imposed price controls on gasoline, the length of the gasoline lines could be expected to adjust to clear markets. In an aggregate model, such adjustments in waiting times could be seen as shifts in the demand functions for gasoline. Such adjustments could be incorporated in the modeling system *if* it were known what shifts might occur.

The success of the equilibrium price-model in predicting politically feasible outcomes depends on relatively large responses on either the supply or the demand side of the market to changes in price. If the responses (elasticities) are both very small, price increases (e.g., in oil) may cause large real- income changes (people cannot reduce their oil bills by buying less), and the public is politically unlikely to let the market work. Thus, the model user must consider political infeasibility of some possible equilibria.

Uncertainty

As discussed previously, forecasting and policy analysis require understanding the degree of uncertainty in model outputs. As discussed in Chapter 2, careful analysis of uncertainty requires a large number of model runs, using not only the standard modules, but also various combinations of alternative modules. Unless some method is used to shorten the time this analysis required, it is so time- consuming in practice that it is seldom conducted.

One requirement for the NEMS, therefore, to analyze uncertainty, is the ability to run the model very quickly. One method of doing so relies on the use of reduced-form versions that can each be run quickly, to rapid system convergence to an equilibrium.

Such analysis would be conducted in two stages. In stage one, the uncertainties associated with each individual module would be characterized by exercizing that module. Two elements of uncertainty are particularly important for projections: uncertainty about the projections based on a particular set of inputs, and uncertainty about actual values of the input variables. These uncertainties could be used to construct a stochastic version of the reduced-form version for a given sector. This simplification could be achieved by imposing a probability distribution on the parameters used in the reduced- form version to correspond with the probability distribution associated with the full module.

Once the stochastic reduced-form versions were developed, the overall system could be configured to run based entirely on reduced-form versions. Monte Carlo simulation techniques could be used to randomly select combinations of the various parameter values for each of the reduced-form versions. Then the entire model could be run until it converged to an equilibrium. This process of randomly selecting parameters and running the overall model could be reiterated until sufficient data were obtained.

The advantage of using reduced-form versions for such a process is primarily that of speed. The greatly reduced turnaround time that the use of reduced-form modules permits may bring such Monte Carlo simulations to manageable proportions. In addition, the use of reduced-form versions may make it easier to incorporate purely subjective probability distributions for the underlying parameters. This ability might encourage researchers to avoid underestimating the uncertainty associated with the underlying models and therefore to avoid underestimating the uncertainty associated with the system as a whole.

Contingent Strategies

As discussed previously, many private and public policies cannot be regarded as predetermined, but rather should be seen as responsive to the immediate, externally imposed situation. For example, were it not for the rapid oil price increases in 1973-74, Presidents Ford and Carter probably would not have initiated the energy policy proposals of that period nor would Congress have passed legislation such as that imposing corporate average fuel economy (CAFE) for new cars. Thus, energy demand and supply functions for energy were themselves shaped by unexpected events. Similarly, the U.S. Synthetic Fuels Corporation could be seen as having pursued contingent strategies in attempting to create energy technologies that could be more rapidly developed by the private or public sector, in case changing energy situations created the need. As it turned out, evidence mounted that energy prices would not rise as rapidly as previously expected and that the costs of synthetic fuels would be higher than some had expected, and the entire effort was disbanded.

Such contingent strategies are conceptually important in modeling the uncertainty in the energy system. In a deterministic forecast, contingent strategies can be treated exactly like deterministic policy choices: they are adopted or not. Thus, for deterministic modeling there is nothing special about contingent strategies.

The essence of contingent strategies is that policy rules themselves may depend on features of the energy system. For example, if energy prices go above a particular point, then energy conservation rules may be imposed. If energy prices remain below another level, or are projected to continue below some level, then such policies may be repealed. In a stochastic modeling environment, contingent strategies will be governed by one set of rules in one state of the world and by another in a different world state.

Projecting such endogenous policy changes is particularly difficult and not always relevant. However, if they can be predicted with some probability, then such contingent strategies can be encoded into NEMS. Within the relevant modules, there would simply be logical statements that test whether the state of the system is such that new rules would be triggered or old rules eliminated. If so, then the relevant equations in the modules would be adjusted. Thus, the conceptual difficulty with analysis of contingent strategies is their clear specification in the first place.

There would also be practical difficulties in addressing contingent strategies. It will be much more difficult, if not impossible, to use reduced-form versions of modules that accommodated contingent strategies. However, reduced-form versions could still be used for other modules. Clearly, the analysis and computing time, but especially computing time, for examining contingent strategies may be extremely large.

The Formation of Expectations

Yet another question, related to that of disequilibrium, is how models should represent expectations of future prices (Muth, 1981a, b; Lucas and Sargent, 1981). Several alternative assumptions are possible. The simplest is "myopic expectations," that individuals in the market assume future prices will remain the same as current prices (Muth, 1981a, b; Lucas and Sargent, 1981). In a stochastic setting, "myopic expectations" are calculated using the random-walk model that has been highly successful in forecasting financial markets.

Alternatively, the "extrapolative expectations" hypothesis assumes that individuals extrapolate recent trends in prices to forecast future prices. Individuals do adjust, but not instantly, to changing informatin, and a random walk characterizes the trend lines of energy variables.

Finally, individuals assumed to have "rational" expectations forecast prices and other economic conditions consistently with the model. When economic actors behave "rationally," rather than "adaptively" they do not gradually change their behavior in response to new information or different circumstances but adopt new decision rules as information or circumstances change. Decision makers are also assumed to look to the future -- to their expectations -- rather than just to the past -- their experiences -- when making decisions. Assuming rational expectations, if the model is deterministic, then individuals forecast prices correctly; if the model is probabilistic, then individuals forecast prices using the same probability distribution about future prices that results from the model.

The assumption of rational expectations emphasizes the important point that people learn from their pasts and try not to repeat mistakes. It reflects the observation that behavior sometimes does change quickly and discontinuously. Because expectations about the future are strongly conditioned by our experience, the real question becomes the extent to which our behavior is rational or adaptive. Learning theory researchers have found that people often make systematic mistakes and that they take time to move from one mode of behavior to another. Thus neither theory perfectly predicts actual human behavior.

There is a further conceptual distinction between rational expectations and adaptive expectations approaches when incorporated into policy relevant models. Rational expectations, or self-consistent expectations, when incorporated in models, embodies the basic concept that government policymakers cannot know the future better than private actors. Use of rational expectations concepts in NEMS would be consistent with a belief that policymakers cannot be assumed to have such privileged access to information about the future. The use of adaptive or myopic expectations implies that government planning can always improve on private decision, because the government can be assumed to know the future better than private actors.

It is currently an open debate as to which assumptions about expectations most accurately, or most inaccurately, describe the world. And a number of energy policy debates turn on disagreements as to whether governmental regulation can improve on private decisionmaking from the perspective of the individual actors. Thus the appropriate description of expectations will not be easily resolved. For that reason it is important that the modeling system be capable of addressing each of the assumptions about expectations formation.

These assumptions can influence model projections including assessments of particular policy options. Unless the process of expectation formation is considered explicitly by NEMS, it will be difficult to assess the system-wide consequences of programs designed to educate people about energy issues or of behavior shifts occasioned by changes in beliefs. If people systematically underestimated future energy prices and therefore systematically bought more energy-intensive capital equipment than they would otherwise, this belief, if identified, might lead to a policy of advertising that energy prices are likely to increase. In a world of myopic expectations, such a strategy might significantly change behavior. In a world of rational expectations, such a strategy would have little or no effect on behavior, since the assumption of rational expectations is contrary to people systematically understating future conditions.

A less obvious example illustrates the feedbacks that may depend on expectation formation. Assume that incentives are proposed to increase the efficiency of newly purchased cars beginning in five years. Such a policy might significantly reduce consumption five or ten years from its implementation. But the consumption reduction might reduce the price of gasoline and thus make it more cost-effective for consumers to buy less fuel-efficient cars in the first few years. If we assume rational expectations, then the model will show that consumers will buy <u>less</u> efficient cars in early years and will therefore consume more gasoline during the first few years the policy is in place. Even over a longer period, the earlier purchases of inefficient cars would increase gasoline use compared to the case if there were no such anticipation of a price drop. If the model assumed myopic or extrapolative expectations, then it would show consumers in the first few years as not anticipating the price reduction and thus not changing their automobile purchases. These models would show that in all years the policy would reduce gasoline use or leave it unchanged. Thus, assumptions about expectations formation can change the magnitude and even the direction of overall impacts of some policy impacts.

If either myopic or extrapolative expectations are believed to best describe world, then NEMS can be run on a year-by-year basis and price data can be based on either the current price or the current and previous prices. However, if rational expectations formation is assumed and a deterministic model used, then NEMS must be solved for all years simultaneously, since supplies in any year will depend on prices projected for future years as well as current and past prices. In principle, iteration based upon all prices would be similar to iteration based upon only the current prices. However, the greater dimensionality of the problem could be expected to increase the time until convergence is obtained, perhaps greatly. In such a case, it would be important to devote attention to computationally efficient algorithms.

For a probabilistic model, however, the process would involve an immensely more complex, time-consuming endeavor. In principle, one would start with a probability distribution of prices, associated with the full range of uncertainty. This distribution would lead to one vector of actual prices for each possible outcome. Finding this one vector would require the iterative processes discussed above, and the process would have to be repeated for each resolution of the uncertainty. The result would be a probability distribution of prices over time. If this distribution were consistent with the starting distribution of prices, then the solution would be a "rational expectations" equilibrium. However, if the distributions were not consistent, then the starting distribution would be adjusted and the process repeated. This sequence would be continued until convergence were reached.

While in principle rational expectations equilibria could be found in this way for a probabilistic model, in practice the process would be so time-consuming as to be entirely impractical unless better algorithms are developed and computer speeds increased above those currently available. Thus, the NEMS will not be used routinely to examine probabilistic rational expectations equilibria, even though rational expectations concepts are logically most consistent only in a stochastic environment. Yet it is important that the model capture as well as possible the effects on behavior of changing information, particularly information changing as a result of contemplated policy actions.

Environmental Constraints

The NEMS is likely to be used to examine the economic, environmental, and security implications of various types of environmental constraints. Historically, the instruments for environmental regulation were primarily technology-specific requirements relating to the use of so-called "best available control technology." Such rules set limitations on the technologies that could be adopted in the regulated sectors. Simulations of the effects of such restrictions could readily be incorporated into the proposed system by modifying descriptions of technologies underlying the various supply and conversion models.

More recently, however, more market-related policies have been adopted, such as the marketable-permit system for acid rain precursors. Such regulations impose system-wide or sector-wide constraints on the amount of residuals emitted. System-wide or sector-wide constraints can also be incorporated into NEMS, by bringing some parts of the environmental impacts module into the integrating system. A per-unit emissions price (e.g., a price per ton of airborne sulfur) could be associated with discharge of the measured residual. Total discharge could be calculated at each iteration based on the economic incentives associated with the estimated emission price. Equilibrium would obtain when the total emissions "supplied" by the system are identical to the maximum emissions allowed. When these two quantities differ, the emission price could be adjusted from iteration to iteration in the same way other prices are adjusted.

Figure 3-5, illustrates the convergence process for system-wide environmental constraints. This figure is similar to Figure 3-2, but in place of the demand module response is the maximum total emission, set by the regulations. The emission "supply" response is represented by the downward sloping curve, which projects total emission supply price for each possible total emission level. This supply price would be based on the estimated behavior of the industries modeled. Equilibrium would be represented by the emission price and emission quantity at which the two lines cross.

In Figure 3-5, an initial price is chosen, P_0, far from the equilibrium price. The maximum total emission remains at Q_{Max}. The modules projecting emissions would then read the price P_0 and provide an emission estimate of Q_0. The integrating system would be at a solution if Q_0 were close enough to Q_{Max}. If Q_0 exceeded Q_{Max}, the next iteration would be based on a higher price, and conversely, if Q_0 were smaller than Q_{Max}, the next iteration would be based on a lower emission price. This new starting point is represented in Figure 3-5 by P_1. The process is repeated until projected and maximum emission are sufficiently close.

OPERATIONAL ISSUES FOR NEMS DEVELOPMENT

Several operational issues should be given careful attention in constructing a modular system. Such attention should be given to the tradeoffs among model detail, needs for data to support the model, computing hardware, and software requirements. It is useful for each module to be scaled to about the same size and detail, so that no single module dominates the choice of computer hardware and limits the usability of the overall system. However, the EIA models include at least one--the coal model--that is far more detailed and larger than useful for the NEMS. Such models will need to be greatly streamlined and simplified. Other models may need expansion to provide the detail appropriate for the system.

Although a modular system could be developed through decentralized teams, in practice full decentralization is not feasible. Rather, constraints should be imposed beyond those implied by the interfaces. For example, the NEMS office may specify that a given module must not require more than a certain amount of comuter memory, to allow the entire system to be operated on a personal computer. EIA personnel could be expected to examine models for quality assurance. Particular coordination would be required to assure that modules all use common data bases, common economic assumptions, and regional, product, and time definitions, as appropriate.

The time sequence of efforts will be important, because it would be unrealistic to expect all NEMS modules to be developed during the first year of NEMS efforts. Perhaps construction of the integrating framework and the control module should be given the highest priority. Early attention could be paid to reducing the size of any unnecessarily large model that would be included within the integrating system. The EIA should build the integrating framework and control module from scratch, after determining the appropriate regional, temporal, and product definitions. However, development of the other modules probably should initially rely heavily upon models already in existence in EIA or elsewhere. Such development might involve bridge or conversion systems to link existing modules into the new integrating framework. It might involve recalibration of modules to the new regional, temporal, and product structure. Or it might involve redevelopment of existing models, using existing conceptual structures, empirical evidence,

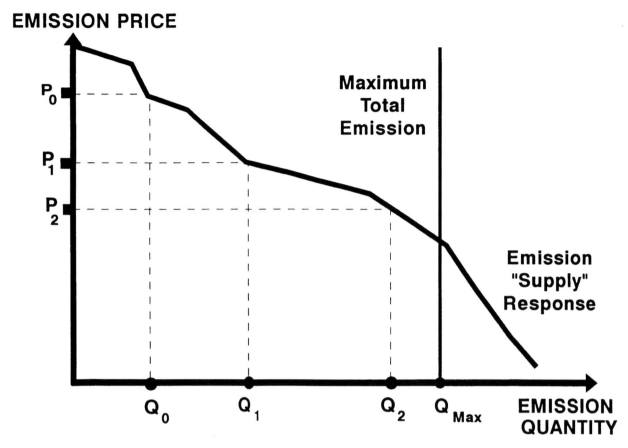

FIGURE 3-5 Modeling a system-wide environmental constraint.

and data, but entirely new computer code. Over time, it is envisioned that new modules would be developed and would supplant these modules.

Several issues are particularly important for redeveloping existing modules. EIA should incorporate the capability of projecting key system-wide environmental impacts into each of the models as they are redeveloped. In particular, emissions of various air pollutants should be estimated for each of the modules within the system.

While modules are being constructed, EIA should give attention to building in to the modules explicit representation of factors that might be modified by specific policy actions. Examples of policy actions that the committee believes to be important include those associated with taxes, environmental constraints, conservation incentives and regulations, and new technology development.

To use reduced-form versions, EIA would have to develop the appropriate methodologies and software for estimating response surfaces. While other researchers have developed methods, EIA personnel or contractors would need to further develop these approaches or to contract for their further development. Because reduced-form versions would greatly facilitate NEMS use for quick-turnaround analyses, the development of the methods and software to estimate response surfaces should be given fairly high priority.

RECOMMENDATIONS

o **The NEMS should be designed to be modular in structure, so to be flexible in its use and to readily accommodate the substitution of alternative models (modules) in the system.**

o **A single model should not be used for short-term, medium-term, and long-term analysis. Similarly, the DOE should not select the model to be used for medium-term analysis based on the desire to conduct long-run analysis through the same framework.**

o **The EIA should attempt to acquire an existing interindustry growth model and should not attempt to develop one itself.**

o **Development of the NEMS can rely extensively on existing models. However, several new models should be developed or acquired from external sources: beyond the interindustry economic growth model, these should include an environmental impacts model and a renewable energy supply and conversion model. Demand models need to be modified to enable a broader range of policy analysis than is possible with the current EIA models.**

o **A reduced-form version of each module should be developed and integrated in to the NEMS. Such versions should approximate the response surface of the full modules. For typical policy analyses, the full module would be used for the sector examined and reduced-form versions would be used for other sectors. Reduced-form versions could also be used in uncertainty analyses, tests of the integrated set of modules, and quick-turnaround policy analyses.**

o **The NEMS should be configured to run on personal computers or workstations, unless such hardware constraints would entail significant loss of the capabilities envisioned for NEMS.**

o **One or more EIA analysts should be charged with the general knowledge of all modules, even those not developed at EIA.**

o **In developing the environmental module, initial efforts should be focused on quantifying direct and indirect air emissions. If resources are not sufficient for this** task, then EIA should focus attention first on the greenhouse gas emissions associated with the extraction, production, transportation, and use of energy.

o The NEMS should capture the effects on behavior of changing information, particularly information changing as a result of contemplated policy actions. Model users should be able to include alternative assumptions about the formation of expectations, including those defined theoretically as myopic, adaptive, and rational expectations.

o The NEMS should provide carefully envisioned graphics and report writers to provide routine graphical and numerical outputs of the types normally found helpful for analytical purposes.

o A major effort is needed to collect more extensive data and information on the U.S. energy system, especially on end use. In such an effort, DOE and EIA should consider the following points:

- Essential to bottom-up demand modeling is knowledge of underlying activities: housing, commercial buildings, industrial production, and transportation. To obtain such information, EIA needs to improve its link to other data-gathering entities.

- Where behavioral information is inadequate, EIA or other DOE offices should help generate interest in obtaining it, by soliciting research, holding workshops, or stimulating other agencies to sponsor research. EIA should devote some resources to sponsoring research in this area.

4

IMPLEMENTATION OF THE NEMS

The previous chapters have considered the use of modeling in the context of energy policy analysis, identified general requirements that a National Energy Modeling System (NEMS) should meet, and defined architecture and data requirements for the NEMS. NEMS will also require the right implementation, management, and resources over the long term to serve its client community. This chapter addresses these NEMS needs.

LEAD ORGANIZATION FOR THE NEMS

The committee recommends that the Energy Information Administration (EIA) develop and manage the NEMS. Congress established the EIA within the Department of Energy (DOE) to serve as an independent, nonadvocacy agency to manage a central, comprehensive, and unified program of energy data collection, analysis, and reporting for the federal government. By statute, EIA provides routine, periodic data reports concerning energy supply, demand, and price forecasts for primary energy resources, fuels, and electricity; forecasts of energy supply and demand as reported in the Annual Energy Outlook; and other analyses. EIA reports are widely disseminated to Congress, the administration, and the public (EIA, 1990b, 1990d). The EIA Administrator must be confirmed by the U.S. Senate.

EIA's current personnel and operational resources provide the needed basic capability and experience to undertake NEMS development. To fulfill its mission, EIA conducts surveys and develops and maintains a variety of models. EIA data and analytic capabilities were used extensively in the recent National Energy Strategy (NES) exercise, and they will certainly be used again in subsequent NES analysis. In the recent exercise, EIA broadened its outreach efforts by collaborating with DOE's Office of Policy, Planning and Analysis, program offices, such as that for Conservation and Renewable Energy, the national laboratories, and with other organizations. This process of outreach should be extended and broadened.

As explained earlier, the committee recommends that NEMS development should begin by building on existing models and data, as appropriate, capitalizing on the best capabilities available to begin configuring the NEMS. NEMS is compatible with EIA's statutory obligations. Thus, EIA can sustain the evolution of the NEMS, expand its own outreach simultaneously, and help NEMS stay current with relevant developments in other agencies and the private sector. The EIA has recently reorganized to accommodate and manage its expanded role with the NEMS. The committee concurs with this reorganization.

It would be much more difficult for DOE's Office of Policy, Planning and Analysis to develop and manage the NEMS on a sustained basis. The Policy Office advances the philosophy and policies of the Department and the administration, and the director of the office serves at the Secretary's pleasure. The Policy Office is, therefore, subject to pressure by advocates from which it would be difficult to shield the NEMS. In the committee's view, the Policy Office should avail itself of NEMS services provided through the EIA, but this reliance does not preclude the office from undertaking some modeling and analytic activities on its own.

The committee believes it desirable that, concurrent with the development of the NEMS, the DOE policy and program offices should continue to develop and use their own more focused modeling and analytical capabilities to pursue their missions. These activities, just as those outside DOE, represent beneficial redundancy. Such capabilities within DOE, however, should not directly duplicate those of the NEMS.

SUGGESTIONS FOR IMPLEMENTATION

The following suggestions are intended to strengthen the internal organization of EIA, enhance its external relationships, and improve EIA's quality of service to DOE and other clients.

The committee recommends that the NEMS be quickly developed so it can be used in preparing the next NES. To achieve this goal and enjoy the significant benefits that would follow, DOE should make a major commitment in human and dollar resources to the NEMS now.

NEMS development requires financial support sufficient for both staff and contract research, so that existing data and models can be further developed and new data and models created. An early EIA project in implementing the NEMS should be the development of multiyear budget and staffing projections. A formal NEMS development schedule should be prepared simultaneously, as the committee suggests later in this chapter ("Management of NEMS Development").

In light of federal fiscal constraints, the committee recommends that DOE first critically review current budget demands for data, modeling, and analysis, both within EIA and throughout DOE, and reallocate financial resources and personnel to help ensure the rapid development of the NEMS. It will be a challenge to reorder priorities, but doing so will produce great rewards in the form of better informed energy policy and planning. The committee is concerned that if the NEMS is not ready to be used in preparing the next NES, other expedient but scattered modeling investments will be made, repeating the experience of the 1991 NES as described in the committee's first advisory report (NRC, 1991a; also see Appendix B).

The EIA Administrator should be directly responsible for developing and introducing the NEMS, because the NEMS will affect most other agency activities, will require new functions, and will overlap relationships with other DOE offices. Therefore, the Secretary should designate the EIA Administrator to be chief executive for the NEMS and make this assignment one of the Administrator's performance requirements.

The EIA should act as the proprietor and custodian of the NEMS, by leading in its creation and introduction, and then maintaining, operating and upgrading it to provide service to its clients. This stewardship will expand EIA's role in serving those who shape policy and strategy. This new function will be in addition to, and must be accomplished in a manner compatible with, the many services EIA provides to present clients.

The EIA must establish the NEMS with the understanding that its efficacy will rest on the quality of NEMS data, models, and analytical capability and also on EIA's responsiveness to clients.

EMPLOYEE ENVIRONMENT

The full potential of the NEMS will be realized sooner and with greater assurance with the creation of an EIA employee environment that attracts and holds highly qualified professionals. It is critical that the architects of NEMS have both a broad view of the kinds of policy issues the system must addresses, and a broad and critical understanding of the merits and demerits of various modeling approaches. Since NEMS incorporates both engineering and economic models, the design team should include people who are comfortable with both. Statistical expertise is needed to ensure that there is a well worked out protocol for incorporating data, updating model parameters, and validating model conclusions. Since principles of software design and database management are involved, there should be considerable expertise in these areas. Accordingly, the following and other complimentary actions should be actively pursued by EIA with the full support of DOE. In the committee's view, this is especially important since the existing DOE/EIA "culture" has suffered after more than a decade of budget stringency and general lack of support from the Executive Branch.

Employees should be encouraged, and appropriate support provided, to advance professionally through such avenues as formal continuing education, credit for authorship of EIA publications, opportunities for applied research, and ongoing participation in national and international forums. Such scholarly activities provide greater credibility to EIA and its professional staff as well as enhancing their professional capabilities.

There should be a regular presence within EIA of visiting professionals who work in collaboration with permanent staff. These professionals would include exceptional individuals on sabbatical, temporary transfers, interns, postgraduates, loaned personnel, consultants, and volunteer advisory groups. Such talents will be attracted if they can actively participate in EIA processes and can carry out activities that will add to their careers.

The EIA should also make full use of the Intergovernmental Personnel Act of 1970 to attract critically needed competencies, including top management personnel. Such talent will be particularly important during the early development and introduction of the NEMS. Properly managed, this infusion of talent can contribute to a better NEMS while improving the esteem of EIA staff, the involved professionals, and the agency.

Employees should be encouraged to see themselves as marketers of vital data, modeling, and analytical services to clients whose needs must be understood through continued communication and satisfied through products of superior quality. To prevent the NEMS from becoming irrelevant, a nonmarketing approach should be strenuously avoided; EIA should not assume it understand its users' needs. The development of an effective NEMS will require the involvement of both model developers and potential users if it is to satisfy these clients' needs.

It should be a goal of EIA in establishing the NEMS to become the national center for energy modeling and analysis. Through this center will flow new ideas, some that will lead to major advances in modeling and analysis. This aim will also create excitement: staff will know their work is significant, contributing to the evolving field of energy policy analysis and modeling.

In addition to enriching the employee's work environment, every reasonable opportunity should be pursued to reward employees monetarily. To meet its objectives, EIA must be able to compete for, attract, and hold competent staff with technical expertise in energy modeling. Such expertise is currently in strong demand by the private sector and current government pay scales are not competitive. Yet the Federal Employees Pay Comparability Act of 1990 provides several avenues for increasing employees' monetary compensation. This law was designed for agencies like EIA, which have major new technical responsibilities, to help them attract necessary talent (EIA, 1991). Provisions of this law allow the following:

- o Recruitment and retention bonuses and retention allowances of up to 25 percent of base pay
- o Hiring above minimum rates for all grades
- o Authorization of up to 800 critical positions with base pay limited to EX - I
- o Waiver of dual compensation restrictions on reemployed civilian and military retirees
- o Advance pay for new employees

o   Travel expenses for candidates and new appointees
o   Higher limits for special pay rates
o   Supervisory differentials of up to 3 percent
o   Time off as an incentive award.

In addition, DOE personnel officials have some authority and flexibility in classifying positions, to help agencies like EIA construct the organizations they need to undertake critical national initiatives.

While other opportunities to improve EIA's employee environment will arise, if the Administrator pursues a combination of those listed above, EIA can attract, retain, and motivate the needed talent and skilled personnel.

MANAGEMENT OF NEMS DEVELOPMENT

An early EIA initiative must be the creation of a comprehensive NEMS development schedule, one that considers to the full extent possible needed modules, models, data and integrations. This schedule must lay out all pertinent activities, events, and relationships among NEMS elements (e.g., management techniques to assess critical paths towards achieving a given goal, such as the critical path method [CPM]). This scheduling will be an early opportunity to involve the affected clients so they can help shape the form and development timing of NEMS modules. Equally important will be early cooperation with the other EIA offices, divisions, and branches whose functions will complement those of NEMS.

EIA must support an active outreach program in which key managers and specialists participate in the best forums within government, academia, and the private sector. Such a program is essential for the EIA to keep abreast of new ideas and techniques. Again, byproducts of these activities will be the greater motivation and credibility of EIA and its staff.

EIA must establish close and meaningful relations with NEMS users and the scientific community. Clients should participate not only in reviewing data needs, model design, and scheduling of model development, but also in setting EIA standards for service turnaround time, quality, and other client needs. The EIA must create adequate forums for such ongoing important communication.

In particular, the committee recommends that EIA form a Users Advisory Council of NEMS clients. The EIA Administrator would determine the council's makeup, which should represent DOE's Office of Policy, Planning and Analysis, and program offices, other federal departments and agencies, and key congressional committees, industry groups, and environmental organizations. This broad, cooperative involvement in NEMS will help ensure its value and credibility. The Users Advisory Council and other forums, such as focused task forces, will contribute to wider understanding and support of analytical results, especially those relating to controversial policy. Without abdicating legal responsibilities, the Administrator is encouraged to rely heavily on the advice of the Users Advisory Council, and to defer to its recommendations whenever possible.

Through use in the next NES and other applications, the NEMS will steadily improve. As experience illuminates NEMS strengths and weaknesses, changes can be made to upgrade the models and improve their ability to inform decision making. In time the NEMS should be used to produce the EIA's Annual Energy Outlook and its other studies and forecasts, which would lead to its continuous use and contribute greatly to the system's value.

Future NEMS clients made it very clear to the committee that timeliness of service is often critical. Ability to provide service with a turnaround time of an hour or hours (including preparation of hard copy) may determine whether the NEMS is useful or not in many situations. The committee also heard presentations on the importance of documentation, quality control, and archiving to protect the integrity, objectivity, and reliability of EIA products and services. Clearly, EIA's internal regulations and policies must be carefully crafted to balance the needs of clients and internal quality control.

The committee suggests that affected parties participate when EIA develops and modifies all standards and criteria for upgrading, developing, maintaining, documenting, and modifying data, models, and analytical services. One of the greatest challenges in developing the NEMS will be promulgating the use of these standards. EIA must protect its statutory independence and still serve many clients. If not anticipated forthrightly, conflicts between the two roles, enforcing internal standards and serving the needs of policy clients, may frustrate achieving the full potential of the NEMS.

Conflicts can be minimized through the early and careful development of standards. An approach to avoid is establishing detailed prescriptive rules that unduly bind employees' use of judgment. Standards for clients would stress service, quality, timeliness, and ease of use. Employees should be free to use their judgment within certain guidelines. Operating within such a framework is far better than a labyrinth of rules that overly restrict employees, depriving clients of the service they expect and deserve.

As discussed in Chapter 2, the DOE's Policy Office will often work in a demanding policy analysis mode. Although the EIA will develop, manage, and maintain the NEMS, the fast-turnaround analysis often required by DOE's policy and program offices can be accomplished with the use of reduced-form models (see Chapter 3). The committee suggests that during early analysis of policy alternatives, model outputs could be generated without concern for the full application of EIA standards, and these outputs could be clearly identified as not carrying the EIA imprimatur.

Standards restricting EIA publication of industry-derived data should not necessarily apply to the NEMS. When the best data for a model are available only from external sources, then these data should be used. In using all data, reasonable care should be taken to preclude biases and inaccuracies. In using non-EIA data, however, special care should be taken both to examine actual bias and to avoid using biased external data.

In particular, EIA standards should encourage the broad use of licensed proprietary models for the NEMS. Restrictions on the use of proprietary models create problems with transparency. Users must be able to run these models on their own computers. Because the best models available should be used in the NEMS and some of the best models available are proprietary, EIA should make every reasonable effort to negotiate provisions for their broad use to achieve the greatest quality and transparency in NEMS modeling. However, the EIA should consider rejecting any licensing agreements that unduly restrict a model's use throughout the entire NEMS system.

In implementing NEMS, EIA will have an interest in both basic and applied research and development (R&D) to improve data gathering, modeling, and analytical methods. While the EIA will occasionally be engaged in applied R&D, the committee does not encourage the EIA to perform basic R&D internally. Instead, when deemed beneficial, EIA should participate in joint basic R&D with those that can help provide funding and participate in oversight and evaluation of the work. Work should also be funded in universities and other high-quality research establishments. Both basic and applied research hold promise for the NEMS, for example, in the areas of uncertainty and long-range modeling and analysis.

MOTIVATION AND USEFULNESS

NEMS will be of little value if not widely used by decision makers. The development of certain relationships is thus critical if NEMS capability is to be integrated in DOE's policy analysis.

The committee recommends that key EIA employees should have direct access to, and involvement with, the decision makers of client organizations, particularly the DOE as needed. If NEMS is to support the development of energy policy successfully, then EIA's top analysts must work closely with decision makers. Equally important, these decision makers must engage EIA analysts in the policy analysis process. This cooperation will stimulate more accurate and comprehensive analyses of policy alternatives. To use NEMS services appropriately, decision makers must have a clear understanding of NEMS's strengths and weaknesses and of how they affect particular analytical efforts.

RECOMMENDATIONS

In summary, the committee makes the following recommendations for NEMS implementation:

o   **The EIA should develop and manage the NEMS, and should move quickly to configure the initial NEMS within the next one to two years and apply it to policy issues including the next National Energy Strategy.**

o   The EIA should move toward using the NEMS as a basis for its <u>Annual Energy Outlook</u> and toward providing modeling support for energy policy analysis more generally.

o   The Secretary of Energy should designate the EIA Administrator as chief executive for implementation of the NEMS and make this assignment one of the performance requirements of the Administrator.

o   EIA should form a Users Advisory Group of likely NEMS users from within the DOE, other government, federal, state, and regional agencies, and private organizations.

o   DOE should capitalize on opportunities offered by the Federal Employees Pay Comparability Act of 1990 to attract and retain highly skilled staff for the development and operation of the NEMS.

o   The NEMS organization should have direct access to, and involvement with, the decision makers both within and outside DOE who are potential NEMS users.

APPENDIX A-1

SCOPE OF WORK

A Review of the Department of Energy's
National Energy Modeling System (NEMS)

by the National Research Council (NRC)

The Department of Energy (DOE) uses a variety of energy and economic models to forecast energy supply and demand. It also uses a variety of more narrowly focussed analytical tools to examine energy policy options. For the purpose of this scope of work, this set of models and analytical tools is called the National Energy Modeling System (NEMS).

THE PROBLEM

The NEMS is the result of many years of development of energy modeling and analysis tools, many of which were developed for different applications and under different assumptions. As such, NEMS is believed to be less than satisfactory in certain areas. For example, NEMS is difficult to keep updated and expensive to use. Various outputs are often difficult to reconcile. Products were not required to interface, but were designed to stand alone. Because different developers were involved, the inner workings of the NEMS are often not easily or fully understood.

Even with these difficulties, however, NEMS comprises the best tools currently identified to deal with our global, national and regional energy modeling and energy analysis needs.

ANALYSIS OF THE PROBLEM

With the goal of evaluating and improving the NEMS, the National Academy of Sciences, National Research Council is requested to review NEMS and advise DOE in the following areas regarding energy modeling and policy analysis tools:

Energy Data: Are the quality and types of data that are currently gathered and available to DOE adequate for NEMS? If not, what additional data should be gathered? If some critical data are not obtainable, what alternatives are available?

Modeling Assumptions: Are the assumptions embodied in models internally consistent and valid? Is it possible to reduce the number of assumptions currently built into the models and the scenario runs?

Energy and Economic Models: What is the availability of other models or data, not currently being utilized in NEMS, that might be useful in improving the NEMS? Are there any gaps in the NEMS in terms of supply/demand feedbacks, consideration of international factors, weaknesses in analyzing certain policy options or other shortcomings?

Role of NEMS in Energy Policymaking: In general, what are both the capabilities and practical limitations of energy forecasting tools? Taking such limitations into account, what are the appropriate uses of forecasting tools in developing the National Energy Strategy (NES)? What other alternatives are available for developing long-range energy policy? How might they be incorporated into future NEMS/NES development?

PROPOSED EFFORT

A 21-month study by NRC is proposed for accomplishing a review of the NEMS modeling and analysis needs, to be conducted in two parts:

First Part of Review

The first part of the review will be completed six months from contract award. NRC will examine the composition of the NEMS and its initial application with respect to a representative slate of analytical tasks. NRC will define and apply criteria to evaluate the capabilities and adequacy of the NEMS for its intended purposes of forecasting and analyzing a variety of energy supply, end-use and environmental issues. In this context, NRC will review, primarily through briefings and presentations from EIA and other DOE sources, NEMS assumptions and methodologies. Particular emphasis will be given to the review of significant components of NEMS as they concern the cost and performance (output) curves for energy supply and end use technologies, environmental releases. NRC will also review the Department's plans for further development of NEMS. NRC will, within six months of contract award, prepare and issue an initial report of its findings. This report, following NRC peer review procedures, will be forwarded to DOE for use in support of the development of the National Energy Strategy.

The report will present findings and conclusions regarding the efficacy of the NEMS in its early configuration, the reasonableness and adequacy of the underlying data, assumptions and methodology, and make recommendations, as appropriate, for near term action by DOE. The report may also outline the steps NRC will take in conducting the second part of the review.

Second Part of Review

In the second part of the review, to be completed 21 months after contract award, NRC will evaluate the NEMS and plans for further development in greater detail. On the strength of this evaluation, the NRC will provide recommendations for improving the models, data and associated analytical tools comprising the system, and for future model development and data collection in order to enhance DOE capabilities for strategic planning. In its review, NRC will address relevant long-range issues and will appraise the robustness and the flexibility of the NEMS to support the DOE policy making process. NRC will examine the adequacy of the models and analytical tools and the quality of data to support relevant policy analyses and development, in all supply and demand sectors, including renewables and in environmental and new technologies areas.

In particular, NRC will examine DOE needs and priorities for the NEMS over the long term. NRC will then identify what system improvements can be made cost effectively and new directions that could be undertaken within the state of the art, given emerging policy issues and other requirements and constraints that will confront DOE in the early 1990's and beyond.

An important part of this review is for NRC to assess the adequacy of current data collection for NEMS and related analyses in terms of its scope, frequency, sample sizes, and other statistical characteristics. An assessment will be made of the role that new data might service in enhancing the capabilities of the NEMS to meet its intended objectives. NRC will make recommendations on data requirements needed to more effectively support the DOE policy-making process and suggest priorities for new data collection and ways that data needs can be met cost effectively.

In its assessment of the modeling activities in DOE, NRC will define and address issues it considers critical in contexts such as the following:

(1) applications history of models in the NEMS; (2) completeness and adequacy of model documentation both for those running the models and those using them; (3) validity of the models in terms of how well the theoretical concepts, data bases, and operations approximate relevant aspects of the real world, including behavioral effects that impact on energy supply and demand; (4) treatment of uncertainties in both model construction and application; (5) integrity of the models in their representation in the computer codes which make them operational; (6) quality of model forecasts, and extent of out-of-system adjustments to outputs needed to account for conceptual and practical limitations of the models; (7) interfaces within the system and the ease with which the models can be maintained and updated; and (8) usability of the system including costs associated with use and upkeep of the system.

In executing the foregoing assessment, it is expected that NRC will rely on existing information and documentation and will not undertake any original model or data validation efforts. In order to assure timely advice enabling effective allocation of DOE resources to the continuing development of the NEMS, NRC may, as appropriate, issue brief advisory reports on findings and recommendations considered significant. The advisory reports will be subject to peer review. NRC will issue, within 21 months of contract award, a peer reviewed final report to the DOE covering their findings and recommendations during the duration of the study.

ANTICIPATED RESULTS

NRC's report, after the first part of the review is completed, will provide informed judgements on the capabilities and adequacy of NEMS for its intended purposes, taking account of its initial applications and will advise the DOE with regard to further NEMS development.

NRC's findings and recommendations at the end of the second part of the review will provide DOE an assessment of NEMS, suggesting the level of confidence appropriate for its analytical products, and guiding its continued development and application. It is expected that NRC's interim advisory reports and final recommendations will enhance DOE's capabilities for strategic planning over the long term.

REPORTING

NRC's findings and recommendations will take the form of the reports described under "Proposed Effort." All reports will be subject to regular NRC report review procedures. Reports will be provided to DOE and to other appropriate individuals and agencies. Briefings will be conducted as necessary.

APPENDIX A-2

COMMITTEE CHARGE

The following charge was outlined in the committee's First Advisory Report:

> The committee will develop a statement of the desirable analytical, modeling, and supporting database capabilities necessary to support the evaluation of U.S. energy policy options. The committee will examine the correspondence between the current and desirable capabilities and will recommend priorities for development of the NEMS. Special attention will be given to the regional and international context for U.S. energy policy development, and to the interaction between energy and environmental policy formulation and analysis.

The more specific objectives of the committee include:

(1) Recommend a process and priorities for on-going modifications to the existing EIA/DOE modeling capabilities to better serve the purposes of planning, forecasting and policy formulation as exemplified by the ongoing NES process, considering data needs, model enhancements and integration. However, detailed technical assessments will not be undertaken of specific models currently in use at DOE and EIA.

(2) Examine whether improved information, both new data series and focused case studies on how energy systems function, would be critical to achieving the recommended modifications and improvements.

(3) Discuss the appropriate role for models in the policy formulation process, the inherent limitations of models, and the specific practical limitations and tradeoffs of data collection and integration in developing more effective modeling capabilities.

(4) Examine broadly the capabilities of relevant public, academic, industrial, regional, state and international interests and make general recommendations of how the activities of these parties can complement and enhance the capabilities of the DOE and EIA.

Appendix B

APPENDIX B

FIRST ADVISORY REPORT

DEVELOPMENT OF THE NATIONAL ENERGY MODELING SYSTEM

COMMITTEE ON THE NATIONAL ENERGY MODELING SYSTEM

Energy Engineering Board
Commission on Engineering and Technical Systems

in cooperation with the

Committee on National Statistics
Commission on Behavioral and Social Sciences and Education

January 30, 1991

NATIONAL RESEARCH COUNCIL
WASHINGTON, D.C. 1991

COMMITTEE ON

NATIONAL ENERGY MODELING SYSTEM

PETER T. JOHNSON, (Chairman), Former Administrator, Bonneville Power Administration, McCall, Idaho
DENNIS J. AIGNER, Dean, Graduate School of Management, University of California, Irvine
DOUGLAS R. BOHI, Director, Energy and Natural Resources Division, Resources for the Future, Washington, D.C.
JAMES H. CALDWELL, Jr., Washington, DC 20016
ESTELLE B. DAGUM, Director, Time Series Research and Analysis Division, STATISTICS CANADA, Ottawa, Ontario, Canada
DANIEL A. DREYFUS, Vice President, Strategic Planning and Analysis, Gas Research Institute, Washington, D.C.
EDWARD L. FLOM, Manager, Industry Analysis & Forecasts, Amoco Corporation, Chicago, IL
DAVID B. GOLDSTEIN, Senior Staff Scientist, Natural Resources Defense Council, San Francisco, CA
LOUIS GORDON, Professor, Department of Mathematics, University of Southern California, Los Angeles, CA
VELLO A. KUUSKRAA, Chairman, ICF Resources Inc., Fairfax, VA
JAMES W. LITCHFIELD, Director of Power Planning, Northwest Power Planning Council, Portland, Oregon 97204
STEPHEN C. PECK, Director, Environment Division, Electric Power Research Institute, Palo Alto, CA
MARC H. ROSS, Professor, Department of Physics, University of Michigan, Ann Arbor, MI
EDWARD S. RUBIN, Professor, Department of Engineering & Public Policy, Carnegie-Mellon University, Pittsburgh, Pennsylvania
JAMES L. SWEENEY, Professor, Department of Engineering-Economic Systems, Terman Engineering Center, Stanford University, Stanford, California
DAVID O. WOOD, Director, Center for Energy Policy Research, Massachusetts Institute of Technology, Sloan School of Management, Cambridge, MA

Staff

MAHADEVAN (DEV) MANI, Study Director, and Acting Director, Energy Engineering Board
ARCHIE L. WOOD, Acting Executive Director, Commission on Enginerring and Technical Systems
JUDY AMRI, Administrative Associate
PHILOMINA MAMMEN, Study Assistant

I. INTRODUCTION

The National Research Council's Committee on the National Energy Modeling System was established at the request of the Department of Energy (DOE) primarily to provide guidance to DOE and the Energy Information Administration (EIA) on the long-term development of a modeling system to support national energy analysis and strategic planning. This report is an interim advisory report that has been prepared by the Committee at the request of the Secretary of Energy.

The Committee held its first meeting in Washington, D.C, on July 31 and August 1, 1990, and has since met again on September 20 and 21, 1990 and November 8 and 9, 1990. At the first two meetings, the Committee was briefed on a preliminary system of existing models configured and applied by DOE and EIA to the analysis underway of National Energy Strategy (NES). That system comprised the 2-tier set of models shown in Figure 1.

The Committee has conducted its work to-date using information provided through presentations from DOE and EIA (See Appendix 1) and from reviews of selected documents (See Bibliography). The Chairman and committee members also met with the Secretary of Energy and his staff on November 6 and 9, 1990 to exchange views on the use of existing models and data in on-going energy strategy analysis at the department and on needs for future capabilities of a National Energy Modeling System (NEMS). Largely on the strength of these presentations and exchanges, the committee has made observations on DOE's use of existing models that is the subject of this advisory report.

The report addresses, in broad terms and at a high level of aggregation, the efficacy of existing models configured by DOE to support the NES activities and the adequacy of the underlying data, assumptions and methodology. It also outlines the Committee's goal in focusing on the long-term development of modeling capabilities at DOE. The committee expects to complete its work and publish a final report in the Fall of 1991.

II. FINDINGS

A. Background

The analytical staff of the Department of Energy (including EIA) has been called upon to support the effort leading to the formulation of a National Energy Strategy (NES). A part of that effort involved the generation of one or more "Reference Case" projections of the U.S. energy situation to the year 2030 along with a set of "excursions" (scenarios) off the reference cases.

The existing modeling capability of the department was used in this effort. The predominant group of models used resided in the Energy Information Administration (EIA) and the DOE Office of Policy, Planning and Analysis (OPPA), although some use was made of other models, such as the ARGUS (Argonne Utility Simulation) model of the Argonne National Laboratory.

Ad hoc interagency staff organizations, particularly the NES Modeling Subgroups, were created to assist in developing assumptions and off-line parameters for the modeling effort (See Figure 1). The cooperation involving the DOE policy and program offices, the EIA, the national laboratories and ad hoc inter-agency staff organizations in developing assumptions and conducting off-line estimations for the NES exercise is to be commended.

In the Committee's view, the foregoing process was constructive. The working group format served to pull together and synthesize information. The interactions and the pooling of expertise were commendable uses of personnel resources.

B. General Observations

There is no comprehensive model or group of integrated models existing within DOE, or probably within the federal government, that has the analytical capability to respond to the long time horizon needs imposed by the NES effort. In the Committee's view, the approach taken by DOE in using available models as appropriate, along with off-line supplemental analysis as necessary, was a rational response to the department's need for expedient support of the NES process. The rough integration of the modeled and off-line intermediate analyses that DOE accomplished through the calibration of the FOSSIL2 model has been a useful way to maintain consistent accounting and reporting of results. (See Note in Figure 1 on the calibration of FOSSIL2.)

The aggregate structure of the models used, however, has significant limitations relative to the analytical results reported by DOE and presented to the Committee. Thus, it is important for the decisionmakers who will be using the results of the NES analyses to appreciate the limited power of the existing set of models used for evaluating policy choices. It would be misleading to assign too much quantitative precision to the results of the model runs or to presume that the models incorporate a great deal of relevant detailed information that can enhance judgments about the future impact of policy choices, particularly impacts beyond a decade or two. In the presentations that were made to the committee, little reference was made per se to the validation of the models used in the current national energy strategy analysis exercise. Policymakers should appreciate the important role of the _a priori_ assumptions and simplifications, and the off-line contributions made by the NES Modeling Subgroups in shaping the excursions or scenarios, which to a great extent dictated the results of model runs. In instances where such working groups play such an important role, DOE ought to consider enhancing the working group format with greater outside participation.

As the Committee understands it, the NES analysis and the development of a National Energy Modeling System were initiated by the DOE Office of Policy, Planning and Analysis, which was also responsible for the choice of excursions or scenarios to be analyzed and for the selection and application of the FOSSIL2 integrating model. EIA has provided extensive analytical support in these activities, but not in its customary role of generating, independently, energy information and forecasts such as the Annual Energy Outlook.

C. Specific Comments

1. <u>Long-Run Modeling and Forecasting</u>

None of the current models used in the NES analysis was developed to generate 40 year projections. It would be misleading to presume that the models incorporate a great deal of information relevant to the long-term implications of the excursions or scenarios analyzed.

The models are deterministic, and are based either on historical data and relationships that may or may not have relevance to situations anticipated in the future, or on external judgments about the course of technological and institutional changes. The variances can be high in the relevant parameters as well as in the structure of relationships within the models. It may well be that uncertainties are so great for forecasts over a 30 to 40 year period that deterministic forecasts cease to be useful over such time frames.

There are issues still to be resolved concerning the internal consistency of the models within the two-tier hierarchy of which they are part. (See The FOSSIL2 Model, below.)

2. <u>The FOSSIL2 Model</u>

In the current NES exercise, the FOSSIL2 model is the integrating vehicle used by DOE to yield long-term forecasts. While the calibration and application of FOSSIL2 may be a useful way to (a) achieve rough integration of the analysis performed with other models, and (b) maintain consistent accounting and reporting of results, the aggregate modeling structure (as depicted in Figure 1) has major limitations relative to the analytical results being obtained from the models.

The FOSSIL2 model itself, according to DOE, is not well understood in the department except by a few people in the Office of Policy, Planning and Analysis. And, the staff at EIA indicated that they too did not have in-depth knowledge of the characteristics and workings of the FOSSIL2 model.

The compatibility of the FOSSIL2 model with the other models is also not well understood by DOE. The influence of off-line inputs to the FOSSIL2 model does not seem to have been adequately studied. For example, the influence of the coal-dominated ARGUS model in determining future fuel choices for electric power production in the FOSSIL2 model could be important and perhaps overly restrictive.

Many of the excursions (from the reference cases) analyzed by DOE with the FOSSIL2 model involved substantial changes in the use of energy technologies, and assumptions on technology availabilities, efficiencies and costs. While the FOSSIL2 model does solve for an energy market equilibrium (both prices and quantities), the excursion analyses entailed no reexamination by DOE of the macroeconomic consequences of the technology changes and other changes assumed. Under these circumstances, the Committee questions the outcomes of the excursions (especially for instance those involving high levels of energy conservation) as being potentially inconsistent with reference case economic assumptions. (See also Adequacy of Data, below.)

The Committee sought but did not obtain clear justification from DOE for the selective use of the FOSSIL2 model for virtually all time horizons considered in the NES analysis i.e, to the exclusion of other available integrating models such as the Intermediate Future Forecasting System (IFFS) used by EIA to generate mid-term forecasts.

3. Adequacy of Data

Much of the energy data pertaining to the NES modeling and analysis are on the supply side of the energy markets. To-date, much less effort has been expended on obtaining demand side data. As a consequence, analyses of the demand side are weak.

Parameter values used in the models are based on very limited data and aggregated information about energy consumption and demand side management. For example, there is little information (at DOE and elsewhere) of the potentials for energy efficiency improvements in the industrial sector. The feasibility of new or alternative energy supply and some end-use technologies appear to rely on estimates made by DOE that are based on limited experience and have a documented history of over-optimism. For instance, specifications made by DOE of technology-derived gains associated with clean coal technologies and methanol are important influences in the analysis, but appear to be speculative.

Energy efficiency improvements were invoked by DOE in some cases and ignored in others. Many efficiency measures seem to have been inadequately studied preparatory to the analysis, and may lead to undue reliance being placed on modeling results.

4. Dealing with Uncertainty

In the analyses to date, it is the Committee's view that little attention has been given to dealing with uncertainties regarding assumptions, input data and results, and to risks posed by an uncertain future. More direct treatment of uncertainties will strengthen the NES analyses.

Ignoring uncertainties could lead to undue reliance on "model outputs" and a tendency to overlook the speculative nature of the projections, particularly over long-term time horizons. In context it is worth emphasizing that, for the models considered in the NES analysis, the Committee is not aware of any basis to assign a greater level of certainty to the differences ("deltas") in outcomes among various excursions than to the outcomes of the individual excursions themselves.

III. CONCLUDING REMARKS

Despite the general limitations of forecasting models and aside from the specific limitations of the models used by DOE in the NES analysis, the Committee believes that models can play a crucial role in enabling informed judgments and decisions to be made in matters of national energy policy. Thus, the committee considers it vital that DOE continue to develop and sustain capabilities for analyzing national energy issues using resources from within the department and from appropriate organizations in both the public and private sectors.

But the models used for energy forecasting and policy analysis have inherent limitations that must be clearly recognized. Indeed, if the future were not malleable there would be little interest and less controversy over the validity of forecasting models in general. In terms of the particular models used in the NES analysis, DOE should take care not to overemphasize the models' capabilities in light of the types of deficiencies observed by the committee. Perhaps, in documenting its current analytical efforts with regard to national energy strategy and communicating findings to the public, DOE should emphasize that the analysis of policy options and combinations thereof were defined and developed drawing on many resources, only one of which was DOE's modeling capabilities.

The decision taken by the Department of Energy to develop and maintain an effective National Energy Modeling System (NEMS) is an important step. The Committee is pleased to participate in that process and to assist DOE in achieving that objective.

The Committee views its charge as follows:

The Committee will develop a statement of the desirable analytical, modeling, and supporting database capabilities necessary to support the evaluation of U.S. energy policy options. The committee will examine the correspondence between the current and desirable capabilities and will recommend priorities for development of the National Energy Modeling System (NEMS). Special attention will be given to the regional and international context for U.S. energy policy development, and to the interaction between energy and environmental policy formulation and analysis.

The more specific objectives of the committee include:

(1) Recommend a process and priorities for on-going modifications to the existing EIA/DOE modeling capabilities to better serve the purposes of planning, forecasting and policy formulation as exemplified by the ongoing NES process, considering data needs, model enhancements and integration. However, detailed technical assessments will not be undertaken of specific models currently in use at DOE and EIA.

(2) Examine whether improved information, both new data series and focused case studies on how energy systems function, would be critical to achieving the recommended modifications and improvements.

(3) Discuss the appropriate role for models in the policy formulation process, the inherent limitations of models, and the specific practical limitations and tradeoffs of data collection and integration in developing more effective modeling capabilities.

(4) Examine broadly the capabilities of relevant public, academic, industrial, regional, state and international interests and make general recommendations of how the activities of these parties can complement and enhance the capabilities of the DOE and EIA.

The execution of this charge will require significant effort by the Committee to better understand the prospective applications for a National Energy Modeling System within DOE and possibly by other users; to evaluate the existing modeling resources that are available for use by DOE either internal to the department or elsewhere; and finally, to propose an approach which will enhance the modeling support for future policy analysis and strategic planning given available resources. This effort will require further collection of information from DOE and other sources, and time for evaluation and deliberation by the Committee.

BIBLIOGRAPHY

Energy Information Administration (EIA). July 2, 1990. *Requirements Analysis For a National Energy Modeling System*. Paper prepared by an Energy Information Administration Working Group. Washington, D.C. U.S. Department of Energy (USDOE).

Energy Information Administration. September 19, 1990. *A Comparison of Requirements with Current Capabilities and Issues in the Design of a New System*. Prepared by an EIA Working Group. Washington, D.C. USDOE.

The AES Corporation, Arlington, VA. July 6, 1990. *An Overview of the FOSSIL2 Model: A Dynamic Long-term Policy Simulation Model of U.S. Energy Supply and Demand*. Prepared for the United States Department of Energy, Office of Policy and Evaluation. DOE Contract No. DE-AC01-89PE79041. Washington, D.C., USDOE.

Energy Information Administration. October 1990. *Improving Technology: Modeling Energy Futures for the National Energy Strategy (Draft)*. EIA. Washington, D.C. U.S. DOE.

U. S. Department of Energy. April 1990. *Interim Report. National Energy Strategy. A Compilation of Public Comments*. DOE/S-0066P. Washington, D.C.: U.S. DOE.

U.S. Department of Energy. July 26, 1989. Statement of Admiral James D. Watkins, Secretary of Energy, before Committee on Energy and Natural Resources, United States Senate, Washington, D.C.: U.S. DOE

Energy Information Administration (EIA). 1989. *Directory of Energy Data Collection Forms: Forms in Use as of October 1989*. DOE/EIA-0249(89). Washington, D.C. U.S. DOE.

Energy Information Administration (EIA). 1990. *Annual Energy Outlook with Long-Term Projections*. DOE/EIA-0383(90). Washington, D.C.: U.S. Government Printing Office.

APPENDIX

DOE and EIA PRESENTATIONS TO THE COMMITTEE ON

THE NATIONAL ENERGY MODELING SYSTEM (NEMS)

July 31 & August 1, 1990

"The National Energy Strategy and the National Energy Modeling System"
Linda G. Stuntz, Deputy Under Secretary, Policy, Planning and Analysis, U.S. Department of Energy (DOE)

"The Development and Operation of the NEMS: An EIA Perspective"
Calvin A. Kent, Administrator, Energy Information Administration (EIA)

"Reference Case for the National Energy Strategy"
Eric Petersen, DOE Office of Policy, Planning and Analysis

"Current Configuration and Applications of the NEMS"
W. Calvin Kilgore, Director, EIA Office of Energy Markets and End Use

"Use of Energy Models and Data Systems at EIA"
Lawrence A. Pettis, Deputy Administrator, EIA

"Requirements Analysis for the NEMS"
C. William Skinner, Technical Assistant to the Administrator, EIA

September 20 & 21, 1990

"Update on NES and Key Policy Issues"
Linda G. Stuntz, Deputy Under Secretary, Policy, Planning and Analysis, DOE

"Context for the Analysis of Key Policy Issues"
   --Descriptions of Options
   --Definition of the "Reference Case"
   --Conceptual basis for integrated analysis
   --Choice of issues to "demonstrate" characteristics
     & capabilities of models applied to the current NES effort
Eric Petersen, DOE Office of Policy, Planning and Analysis

"Sectoral Energy Demand: PC-AEO Models"
John D. Pearson, Director, Energy Analysis & Forecasting Division, EIA Office of Energy Markets and End Use

"Energy Supply - Coal & Electricity: NCM & ARGUS"
Mary J. Hutzler, Director, Electric Power Division, EIA Office of Coal, Nuclear, Electric and Alternate Fuels

"Energy Supply - Oil & Gas" GAMS & PROLOG"
Susan Shaw, Analysis and Forecasting Branch, Reserves and Natural Gas Division, EIA Office of Oil and Gas.

"Integration Model: FOSSIL 2"
Roger Nail, AES Corporation, Arlington, Virginia

"Oil Market Simulation Model"
Erik Kreil, International/Contingency Information Division,
EIA Office of Energy Markets and End Use

"DRI Macroeconomic Model"
Ronald Earley, Economics and Statistics Division, EIA Office of Energy Markets and End Use

"Wrap-up on Current NES Analysis Effort"
Robert C. Marlay, Acting Director, DOE Office of Program Review and Analysis

"Update on NEMS Development/Look-Ahead"
Calvin A. Kent, Administrator, EIA

APPENDIX C

THE MISSION AND FUNCTIONS OF THE DEPARTMENT OF ENERGY

ESTABLISHMENT OF THE DEPARTMENT OF ENERGY

The U.S. Department of Energy (DOE) was estabished in 1977 by the Department of Energy Organization Act, which primarily consolidated a number of pre-existing federal agencies and authorities. The creation of a DOE was recommended by the Administration and strongly supported in both Houses of the Congress. The stated purpose of the reorganization was to secure effective management of energy programs, to assure a coordinated national energy policy, and to create and implement a comprehensive energy conservation strategy. (P.L. 95-91, [42 USC1/71013])

The principal building blocks that were assembled within the Department were:

o The Federal Energy Administration, a high level planning and administrative organization that previously had been created to administer the federal authorities stemming from the legislative and administrative responses to the "energy crisis" of the early 1970s;

o The Energy Research and Development Administration (ERDA), under which the energy research and development (R&D) programs and nuclear weapons programs of the Atomic Energy Commission (AEC), other pre-existing, non-nuclear, energy R&D programs, and greatly expanded R&D activities created by new legislation had already been consolidated;

o The federal electric power marketing agencies of the Department of Interior, two of which rank among the largest electric utilities in the nation; and

o Important regulatory functions relating to the leasing of energy resources on the federally administered public lands.

A variety of other, less sweeping responsibilities and programs relating to energy were transferred to DOE from among the other departments. The formerly independent Federal Power Commission was reconstituted within the DOE as the Federal Energy Regulatory Commission (FERC). The FERC still operates as an independent regulatory agency, but the Secretary of Energy has certain coordination and advisory authorities regarding its activities.

The Energy Information Administration (EIA), which was established within the DOE, was charged with a very broad responsibility for

> "...carrying out a central, comprehensive, and unified energy data and information program which will collect, evaluate, assemble, analyze, and disseminate data and information which is relevant to energy resource reserves, energy production, demand, and technology, and related economic and statistical information, or which is relevant to the adequacy of energy resources to meet demands in the near and longer term future for the Nation's economic and social needs."

A number of special provisions were incorporated in the EIA's enabling legislation that were intended to ensure the objectivity, validity, and independence from political bias of the Administration's data and analytical results. The Administrator is required to have a professional background that qualifies him to manage an energy information system, the Secretary initially was statutorily required to delegate to the Administrator certain legislative authorities for data collection (this provision was later amended), and an annual professional audit of EIA's statistical performance is required.

The most interesting of the provisions relating to the EIA is that the Administrator is expressly granted authority to collect information, conduct analysis, and to publish reports without prior approval of "any other officer or employee of the United States with respect to the substance of any statistical or forecasting technical reports which he has prepared in accordance with law."

In practice, of course, there are many practical constraints on such independence, not the least of which are the budgetary controls over EIA's spending. Nevertheless, the intention of the Congress, with the approval of the President upon enactment of the legislation, is clear. The EIA is expected to be a center of valid, unbiased information describing the energy situation, including objective forecasts of energy trends.

EIA also has the responsibility under the Act to furnish information and analysis to the other components of the DOE organization and to make its information and analysis available to the public subject to certain confidentiality restrictions of law. The Secretary also has the authority to utilize the capabilities of EIA in support of other Departmental functions.

As it is presently constituted, the DOE encompasses the major part of direct federal programs and regulatory authorities that impact upon the national energy system specifically for purposes of governmental energy policies and goals. The DOE, moreover, is the governmental agency with the responsibility for surveillance of the energy system to identify public policy issues arising in the evolution of the system and, presumably, with the responsibility to make informed inputs about the energy-related consequences of federal initiatives taken for non-energy policy purposes, as examples, tax and environmental policies. It is also charged directly with implementing conservation and efficiency strategies. There are several aspects of the role of DOE that are directly related to the uses that may be made of a National Energy Modeling System.

THE SECRETARY OF ENERGY AS CABINET SPOKESMAN

In his capacity as a member of the Cabinet, the Secretary of Energy has the responsibility to bring a knowledge of the energy situation to the formulation of overall domestic and foreign policy initiatives. He ought to perceive emerging energy-related issues and propose policies to address them. He should be the source of informed judgment about the consequences for the energy system of policy proposals that are advanced for other social and economic purposes. For example, the Secretary has been explicitly charged by the Congress with a responsibility to provide "independent technical advice to the President on international negotiations involving energy resources."

As a member of the Cabinet, the Secretary also will be called upon to be the spokesman for the Administration on issues and policy initiatives that have important energy-related implications. He will be expected to have an in-depth technical appreciation of the way in which policy decisions will impact energy industries, energy consumers, and environmental quality.

An important requirement upon DOE is to provide the information base and analytical support for the Secretary to comprehend the energy situation, formulate policy initiatives, particularly those that effectuate conservation and incorporate national environmental protection goals, and respond with expert advice to the policy proposals advanced by Executive officials in the Executive Branch and by the Congress.

THE DEPARTMENT OF ENERGY AS STRATEGIC PLANNER

The creation of DOE was responsive to a deeply felt need for comprehensive government surveillance of the national energy system and for the articulation of a comprehensive energy policy. This concept goes beyond the simple consolidation of direct federally sponsored energy programs and regulatory activities and even beyond the government-wide coordination of federal and state activities affecting the energy system. It implies a notion of strategic planning to ensure a secure energy supply for the future as well as "create and implement a comprehensive energy conservation strategy that will receive the highest priority in the national energy program" (42 USC7112[4]). The DOE enabling act also specifies that the Department's role is:

> "to provide for a mechanism through which a coordinated national energy policy can be formulated and implemented to deal with the short-, mid-, and long-term energy problems of the Nation; and to develop plans and programs for dealing with domestic energy production and import shortages."

Despite the wide variation among political convictions about the appropriate level of governmental intervention into the marketplace, the notion that a "national energy policy" is an appropriate goal has persisted. The Congress, in the enabling Act, mandated the DOE to submit a biennial "National Energy Plan" which constitutes a status report on the national energy situation along with forecasts of supply and demand trends. The Plan also was intended to include "the strategies that should be followed and resources that should be committed by government to achieve the welfare and economic objectives served by the energy system, and recommended governmental initiatives to implement the strategies." The biennial National Energy Plan process has been continued with more or less enthusiasm on the part of the Secretarys, depending upon the contemporary attitudes toward the energy situation. The commentary of opinion-makers in and out of government since the advent of the invasion of Kuwait by Iraq, however, confirms that the energy planning process still falls short of expectations and also that the notion of a generic national energy policy remains a widely held goal.

The Bush Administration, under the leadership of the current Secretary of Energy, has revitalized that notion in the form of the National Energy Strategy (NES) process. According to President Bush, the objective of the NES should be:

> "achieving balance among our increasing need for energy at reasonable prices, our commitment to a safer, healthier environment, our determination to maintain an economy second to none, and our goal to reduce dependence by ourselves and our friends and allies on potentially unreliable energy suppliers."

A "First Edition" of the NES was published by DOE in February of 1991 (DOE, 1991a). It followed DOE's solicitation and receipt of extensive comments invited through public hearings and written submittals from a broad range of spokesmen of the energy industries, energy consumers, representatives of all levels of government, and related interest groups. It also involved a substantial, indeed unprecedented, analytical effort carried out through a cooperative effort within the DOE.

The expressed intention is that the NES process will continue. A report is anticipated biennially that will review the energy situation and outlook and will also propose policy initiatives for governmental action to implement the strategy. In its capacity as the strategic planning agency for energy policy, DOE also becomes the focal point for the federal government's ongoing interaction with the energy industries. Through regulatory and data collection functions and cooperative R&D contracts, DOE has day-to-day activities that involve contacts with most of the energy sectors. The Department's policymaking and spokesman roles, along with its technical understanding of the energy system, also make it a natural point of contact for energy industry associations and leaders who wish to influence the direction of federal policy.

THE DEPARTMENT OF ENERGY'S INFORMATION ROLE

As the Executive Branch and the Congress struggled with the critical public policy decisions required by the energy situation in the early 1970s, severe shortcomings became evident in the information available to help formulate and justify policy initiatives. At that time, the federal government compiled very little data relating to the general surveillance of the energy situation. Energy-related items were captured in the general census and economic data collection, but even that meager material was not well organized to respond to the policy questions being asked. Some specific information was available concerning the regulated energy utilities and the activities of the energy industries that pursued energy resource development on the public lands, but it did not provide comprehensive descriptions of even those energy activities.

The energy supply industries themselves had some comprehension of the situation, but even the largest companies had remarkably little knowledge of the overall operation of their own industry sector beyond their own corporate activities and even less knowledge of their sector's interactions with those of the other energy forms, or with end-users.

The information that policymakers derived from industry sources without verification, moreover, was often suspected to be self-serving. In the nearly paranoiac public response to energy price spikes and supply shortages, the policymakers were reluctant to rely upon industry information in making decisions. The industry data also held little credibility with the public when the ultimate decisions had to be justified in the political debate. Many of the most memorable errors in policymaking of the energy crisis era can probably be credited to a lack of comprehension of the practical limitations and capabilities of the energy system.

The DOE was charged with improving the "effectiveness and objectivity of a central energy data collection and analysis program." The central responsibility to develop and maintain a comprehensive description of the national energy situation was vested in the EIA. From the outset, that description was clearly expected to couple valid historical data series with objective and thoughtful forecasts of the future trends.

THE DEPARTMENT OF ENERGY AS R&D MANAGER

Finally, the DOE acquired the energy R&D programs that previously had been consolidated within the ERDA. The initial program emphasis, growing out of the extensive R&D activities of the Atomic Energy Commission, included a heavy emphasis on nuclear energy, and this nuclear-electric orientation persists today. Another major area of emphasis has been basic research, administered largely through the national laboratory establishment. These programs are loosely related to energy and nuclear weapons technologies and include high energy physics, medical sciences, and fundamental materials and process sciences.

During the mid-1970s, however, a series of new grants of authority and appropriations of large amounts of funding created a broad R&D program extending to nearly every aspect of energy supply and demand technology. The DOE energy R&D programs carried out directly and through the national laboratory establishment now encompass basic energy sciences, high energy physics, fusion energy, and the more applied research into the nuclear, fossil, conservation, and renewable energy technologies. In recent years, budgetary constraints along with an explicit policy promulgated by the Reagan Administration to restrict DOE research to long-term, high-risk technologies have acted to bias the R&D program toward the basic sciences and long-range nuclear technologies. Some exceptions, such as the "clean coal" demonstration program, enjoyed Congressional support and have survived. More recent policy expressions appear to signal the reconsideration of that policy and a moderate shift of program resources to nearer-term R&D sectors. Pressures to address global warming issues, in part through increased emphasis on energy efficiency and renewables may accelerate that shift (NAS, 1991). However, large budgetary increases remain unlikely.

APPENDIX D

ILLUSTRATIVE CASE STUDIES

As the committee deliberated the requirements of the proposed National Energy Modeling System (NEMS), it decided to undertake three hypothetical case studies that would generate some generic understanding of how a NEMS might be applied and what kinds of analyses it could provide. The three examples presented in this appendix address energy demand, energy supply and research and development (R&D) program planning. Case Study 1 relates to automobile fuel economy standards, Case Study 2 to natural gas pipeline certification, and Case Study 3 to R&D program planning for magnetically levitated trains. They resulted in some general insights about what a NEMS could and could not be expected to provide as explained later.

CASE STUDY 1: AUTOMOBILE EFFICIENCY STANDARDS

Establishing the Need for Policy Intervention

In our economic and political system it is generally presumed that individuals make economic choices that serve their self-interest, and will balance the costs and benefits of energy efficiency in making decisions such as automobile purchases. The aggregate of all such choices by consumers will provide the market's preference for energy efficiency and guidance to auto producers about the array of options to offer consumers.

This model of perfectly rational decision makers operating in a textbook market often fails to describe the real world. The extent to which the perfect market paradigm characterizes real markets for energy supply or energy demand has a great deal of relevance to the issue of government intervention. If the market is functioning perfectly, and all costs are reflected in the marketplace, then government intervention will always reduce economic efficiency. But if the market fails to perform as microeconomic theory would hold, or if significant costs or benefits are not reflected in market prices for energy or energy services, then carefully designed government policies can improve economic efficiency.

Thus, a necessary condition for government intervention in the market is the existence of significant market failures or externalities. These can be identified either empirically--through studies showing that the market does not produce the results consistent with economic theory--or preferably by identifying specific mechanisms of market failure.

Next, the sufficient condition for government intervention is that the benefits of intervention must exceed its direct and indirect costs. In the case of standards, not only should the benefits exceed the costs, but the benefits of a particular level of standards should be higher than alternative levels, and the benefits of standards as a policy must be greater than the benefits of alternative policies that could achieve similar results.

The quantitative analysis of market failures, and the cost-benefit analysis of alternative policies intended to address those failures, would make use of the National Energy Modeling System (NEMS).

Analytical Requirements

The first step is to analyze market failures that may be affecting the choice of automotive fuel economy. The existence of market failures would explain why there are societally cost-effective improvements in energy efficiency that are not currently being exploited.

The analysis would attempt to identify all possible externalities and market failures with a bearing on the efficiency issue, and determine the best ways of rectifying these failures. This analysis will develop the arguments as to whether government mandated efficiency standards will best address those market failures. Efficiency standards may represent a "second-best" approach to these market failures, in which case the infeasibility of first-best approaches must be established. Plausible externalities to be examined include:

1. Presence of market power over the world oil price;

2. Degradation of environmental resources that are not factored into private consumption and production decisions;

3. The public good aspects of investment in research and development (R&D) that would cause the private sector to underinvest in more efficient technology;

4. Possible differences between social and private discount rates; and

5. Economic costs of oil market disruptions that are not internalized in private decisions (i.e., energy security externalities).

Once the linkage between market failure and efficiency standards has been established conceptually, it is time to turn to NEMS for quantitative analysis that will help set a standard and determine how it is to be implemented. Presumably, the standard will be set on the basis of a quantitative estimate of the magnitude of the distortion in consumer preference for efficiency caused by market failures. Thus, NEMS must include certain ingredients:

1. Data that describes the current vehicle fleet;

2. A model that describes the vintaging of the fleet under different assumptions;

3. A production cost model that calculates the cost of standards and their effect on producer surplus;

4. A consumer choice model that will estimate the effect on auto demand caused by changes in efficiency and auto prices, gasoline prices, safety, and other relevant variables; and;

5. A travel demand model that will estimate the effect of standards on fuel consumption and, together with (d), the effect on consumer surplus.

A more complete policy analysis requires comparisons between the economic costs and benefits of standards with alternative policies that will achieve similar results, such as:

1. Higher excise taxes on gas guzzlers;

2. Rebates on high miles-per-gallon cars, possibly on a sliding scale and in conjunction with (1);

3. Sliding scale subsidies for high mileage cars, funded by gasoline taxes or registration fees;

4. Higher gasoline taxes;

5. Subsidies to consumers or manufacturers for higher efficiency cars; and

6. Subsidies or taxes to promote the use of alternative fuels.

Each of these policies will have implications for the economy, for the environment, and for energy security. Consequently, the analysis must include an evaluation of relationships between each policy action and the primary measures of output discussed in Chapter 2.

Even the most complete form of NEMS cannot accomplish all the tasks described above. Independent analytical input is required in identifying market failures and in establishing a cost-benefit framework that gives meaning to the quantitative output of the models. In addition, key decisions must be made by policymakers in setting the standard (i.e., in balancing costs and benefits), in choosing between standards and alternative policies, and in choosing an implementation strategy.

Nevertheless, it may be concluded that:

o The structure and use of the NEMS can be changed to handle these analyses.

o Not all of the separate pieces of the analysis need be done directly within the NEMS. However, NEMS can provide a consistent framework for the off-line analyses.

o The NEMS will require more and greater data disaggregation on the demand side than currently available.

o The NEMS must provide a better representation of economic decision-making by energy consumers and by other related elements of the economy.

CASE STUDY 2: NATURAL GAS PIPELINE CERTIFICATION

Establishing the Need for Policy Intervention

This problem arose in the following context. The Federal Energy Regulatory Commission (FERC), a quasi-independent regulatory body with commissioners appointed by the Executive Branch, but statutorily independent from the Department of Energy, is charged with the responsibility of certifying the need for new natural gas pipelines. Such a certificate is required before any new interstate transportation system for natural gas can be operated. There are those who believe that the FERC has been slow in issuing these certificates and thus has inhibited the development of new natural gas reserves and the effective use of existing reserves to the detriment of sound National Energy Strategy. Could a model be used to quantify the consequences of inaction by the FERC and thus support the policy prescription that the FERC be abolished and the function transferred to the Department of Energy?

Analytical Requirements

The natural gas transportation problem involves distributing from supply sources to points of end use. As supply will include gas from new fields in the Rocky Mountains and also from Canadian reserves and will include new uses on both coasts, the need for new pipelines is clear. Producers in the new fields are interested in seeing adequate transportation for development of their reserves. Customers are interested in seeing new supplies and in generating gas on gas competition from different regions as a means of holding down prices. Prospective investors in new pipelines see the opportunity to generate profits by matching supply and demand. Several groups of investors are competing for the rights to construct new facilities. Construction of a new pipeline confers at least a partial natural monopoly to its owners. All of the competing proposals cannot be built and operated profitably. Which of the new proposals are in the "public interest?" At the same time, existing producers and pipeline owners have competing proposals for expansion of existing systems as a "cheaper" method of satisfying the need. Should these proposals be accepted as well? What are the consequences of "overbuilding" or "underbuilding?" To the extent that pipelines to the East are constructed and reserves are dedicated to this region, fewer reserves are available to be dedicated to existing or potential customers in the West.

Some of these customers may not have "acceptable" alternative energy supplies. Which allocation of resources is in the "national interest?" What will the consequences for energy security and/or trade matters be if the nation or specific regions become dependent on Canadian gas? These are the types of questions facing the FERC in a pipeline certification case.

Clearly, a model of the natural gas system indicating supplies, the existing distribution system, and points of end use would be useful in answering these questions. In fact, of course, these models exist. The American Gas Association model and the models used by the Gas Research Institute are examples of relatively comprehensive models of natural gas supply and demand that are continually updated and validated in an open forum by competent modelers backed by a comprehensive data base and adequate budgets. Many other such models exist and are operated by public agencies, private interest groups and third party modelers and consultants. There is no doubt that the interested parties to the FERC proceedings have their own models and have used these models to show the benefits of their own proposal or position. There is no doubt that the answers conflict. The answers conflict because of varying model constructs, different data base information, different assumptions about future conditions, and different objective functions. In this setting, how is one to judge the performance of the FERC?

It is certainly possible to use a model to "quantify" a prejudice or preconceived notion one has on the question. Armed with a historical data base and an appropriate model, one could simulate the past twenty or thirty years of energy supply and demand with the only independent variable being new gas pipeline construction. Then using some objective function related to, say, minimizing the cost of energy services, compare the current actual pipeline configuration with the "optimum" configuration computed by the model. All else being equal, the difference could be ascribed to be the "cost" of the FERC. Alternately, using a consistent future scenario, the model could be run with and without new pipeline capacity, or with a "regulatory lag" of an extra, say, two years. Differences in the objective function could be ascribed to inaction or slow reaction by the FERC.

The Role of Analysis

There are obvious problems with either of these approaches even when the model adequately addresses the physical situation. The first problem relates to the "all else being equal" constraint. This constraint is never absolutely true. The more significant new pipeline construction is to overall energy supply and demand, the more difficult it is to make appropriate simplifying assumptions without distorting the answer. At the other extreme, that is new pipeline construction does not matter to aggregate supply and demand, the small difference between two large numbers generated as "the answer" by the model is mathematically extremely imprecise.

The second set of problems relates to the definition of the objective function. The model can be fairly simple and the data required readily available if the objective function is narrowly defined. If cost of service tariff minimization on the pipeline system as a whole over a ten or twenty year period is the only objective, the answer can be obtained without too much difficulty. Natural gas burner tip price minimization with fuel switching requires larger models and significantly more data. As soon as real world considerations of "externalities" including environmental factors in production, transportation and consumption; treaties and trade policy with other sovereign countries; regional political considerations, etc. are included, the problems become very large.

The third set of problems relates to ascribing any difference between the objective functions to a policy solution that abolishes the FERC. Just because the FERC under past commissioners and policies may have "made a mistake," is no reason to expect that the errors will continue or that some alternate regulatory body or the "free market" can do a better job in the future.

The sum total of these problems could dwarf the technical modeling issues and data requirements. In fact, no simple model run is likely to be very useful in grading the performance of the FERC or guiding an alternate regulatory construct. This is not to say that models are not useful in pipeline certification cases. The opposite is true. There is great value in having all parties argue before the FERC or any subsequent regulatory body using consistent historical data, common future reference cases and similar modeling tools. The process will select those models and the data that are most useful to decision making. The process will ensure that data bases are kept current and complete, that new modeling techniques are employed and that externalities are quantified and considered as society muddles through the tradeoffs.

This analytical framework, although developed and used by FERC for pipeline certification, is valuable to a NEMS. Models and data could be transferred wholesale to NEMS or simplified for inclusion in the more general model. At the same time, the consistency of having to use a NEMS data and policy set as a base case for pipeline certification is just as useful in the other direction. However, the question of whether the FERC should be restructured or report directly into the Executive Branch as an advocate of policy is not likely to be settled by a natural gas supply and demand model specifically or a NEMS generally.

In summary, the NEMS should be capable of evaluating the economic, environmental, and energy security implications of the various U.S. natural gas alternatives. The gas model needs to include gas transportation alternatives and be able to deal with the uncertainty associated with changing sources and uses for natural gas.

CASE STUDY 3: MAGLEV R&D

Establishing the Need for Policy Intervention

Evaluating different R&D opportunities at the Department of Energy is one of the more difficult problems for energy models to handle. Nevertheless, the committee believes that modelling can play a useful role in helping set R&D priorities by subjecting all proposals to the discipline of self-consistency with the nation's energy plans and forecasts. Modelling can also help by establishing a uniform context for all projects instead of program-specific assumptions for performing benefit cost analysis. In some cases, merely structuring the decision to involve a benefit cost equation will be a step forward that modelling can help achieve.

While a national energy modelling system can be useful in helping to evaluate R&D projects, it cannot provide a complete answer, because many of the calculations that are necessary--even energy modelling calculations--will require additional detail that is not appropriate to put into the NEMS. NEMS modules should be designed to accommodate this added detail, but the detail should not be added to the NEMS unless the need for this detail is demonstrated on a repetitive basis. Nevertheless, NEMS may have a role in the analysis even where off-line analyses are also needed, to establish consistency. The consistency should work in two directions: the results of more detailed modeling for specific program evaluation should be consistent with the more simplified modeling that goes into NEMS.

Because of this need for interaction between the NEMS and more detailed models, DOE program offices must have independent analytical capabilities.

These conclusions were guided by the following analysis of the case study of evaluating R&D proposals for magnetically levitated passenger railroads (MAGLEV). This case study addresses the questions of what the NEMS should be expected to offer to the R&D portfolio planning process by analyzing what it could offer to the evaluation of high speed rail.

Analytical Requirements

EIA is under pressure to make technology assumptions in projections explicit in order to assist DOE in evaluating the potential benefits and other impacts of candidate programs for R&D funding. The ultimate objective would be to establish R&D program priorities on a benefit to cost basis to assist in program planning and budgetary tradeoffs.

There are several analytical requirements in R&D program planning. NEMS can contribute to some of them:

1. The R&D technical and economic targets must be established for the technology and constantly updated as the R&D progresses. They include target efficiencies and capabilities, target capital and operating costs, target dates for market availability. These are the responsibility of the program manager.

2. The setting or outlook in which the technology must enter the market should be uniformly established to impose discipline on benefit-cost evaluations and comparisons among competitive R&D proposals. A NEMS "base case" without DOE sponsored R&D could provide consistent inputs such as energy service requirements--fuel and electricity price tracks--economic parameters such as inflation rates, cost of money. (There may be more than one such scenario measures of uncertainty.) The program manager is a potential NEMS user.

3. Specific considerations of market entry and competitive technologies to the candidate R&D should be proposed by the program manager and validated in negotiation with NEMS (to the advantage of both). One implication of this is that the program office must acquire substantial analytical capabilities. For this kind of program office analysis to be effective, DOE must invest in research on non-technology issues (e.g., economic market and decision-making and policy issues related to land-use decisionmaking-policy issues) relevant to such analysis.

The Analytic Process and the Use of NEMS

The process of analyzing an R&D option such as MAGLEV could begin with the use of the NEMS base case without advanced inter-city transportation technology research. This will provide a projected description of:

1. Costs of fuels and electricity;

2. Economic parameters--cost of money, disposable income, etc.;

3. Transportation efficiencies--air, rail, vehicle fleets; and

4. Needs for personal transportation. These needs would be contained at some level of disaggregation in the transport module of the NEMS. Possible disaggregations include separation into urban, suburban, and inner city travel, separation of inner city travel by trip distance or by region or by transportation corridor, and possibly down to the consolidated metropolitan statistical area (CMSA) level. To the extent that these disaggregations are not incorporated in the NEMS, they are likely to be necessary as described below for inclusion in a more detailed model that can be used for evaluating MAGLEV.

The program office will then provide:

1. Technological capability of new technology, speed, capacity, electric efficiency (energy use per passenger mile);

2. Cost of new technology--capital, operating;

3. Costs of competing technologies, including conventional technology such as highways, cars, airports, and planes;

4. Estimated market entry date;

5. Probability of success (used to discount potential benefits in comparative analysis);

6. Estimates or models of market penetrations of the new technology, if successful. Outputs should be made compatible with NEMS.

The analytic process would proceed as follows:

1. The program office may use detailed transportation models to evaluate market entry. Parameters would be consistent with NEMS base case.

2. Program office inputs would be run in NEMS to evaluate macro impacts:
    - National Security (oil savings)
    - Environmental impacts (emissions tradeoffs: vehicles vs. electric power production)
    - Macroeconomic impacts (consumer surplus)

Capital requirements of the changed transportation system would be provided by these program managers. These would be incorporated into an economic growth model as part of NEMS (if part of NEMS). That would require a suitable economic growth model to be developed and made a module of NEMS.

3. More detailed model and off-line analysis compare costs of R&D to benefits for comparison with other candidates. Validation with NEMS run reveals contradictions and can lead to negotiations. This more detailed analysis off-line should consider a number of issues regarding MAGLEV including its benefits or disbenefits in convenience relative to competing modes, including conventional rail, European or Japanese style high-speed rail, automobile, and air travel. Such discussions would include dimensions of analyzing which corridors MAGLEV would be appropriate in, whether MAGLEV would have capacity advantages within that corridor compared to other options, whether capacity expansions of airports or highways would be needed without MAGLEV in that corridor and the relative capital costs of those options, and related land-use issues. The process of validation and negotiation would serve two functions: it would provide discipline to the project manager in projecting a stream of benefits from the project consistent with the rest of the nation's energy (and in this case transportation) systems, and may improve the structure of NEMS by comparison with the more detailed off-line model.

4. Over time, technology paths in base case and NEMS at large are improved through negotiations and external program office micro analyses.

The Role of Analysis

1. **Establishment of appropriate base cases.** R&D programs should be compared to a base case of DOE's current R&D budget, which may be zero for a particular technology. One option that must be considered is that entities other than DOE will do the R&D; for example, in this case, private transportation companies or foreign companies or governments. One should also consider the alternative case in which no R&D is done but similar options such as conventional rail high-speed trains using current or incremental improvements on current technologies are built, or that foreign or industrial R&D efforts will produce the needed technologies.

2. **Varying levels of DOE effort.** How much would the proposed program be slowed down if its budget were reduced? Is there a minimum viable level of effort to make the R&D worthwhile?

3. **Double counting alternative R&D approaches.** For example, will advanced vehicle R&D reduce the number of gallons of gasoline that can be saved by MAGLEV?

4. **Alternative NEMS scenarios.** Will the value of MAGLEV change radically under different economic, technical, or policy scenarios?

SUMMARY

In summary, then, these case studies illustrate the following points:

o Important policy options including R&D portfolio planning should be subjected to consistent scenario discipline.

o NEMS can be of value in establishing parameters for this consistency.

o Program offices must have independent analytical capabilities.

o NEMS should integrate additional detail only as the repetitive need is demonstrated, but it should be improved to remain consistent with detailed analysis once it is validated.

o NEMS modules should be designed to accommodate added detail, but all detail must not be added to NEMS and NEMS may have a role in analysis where off-line analyses are also needed.

APPENDIX E

A BRIEF DESCRIPTION OF DOE AND EIA MODELS[1]

PROJECT INDEPENDENCE EVALUATION SYSTEM

The Project Independence Evaluation System (PIES), later named the Midterm Energy Forecasting System (MEFS), was initially developed in 1974 by the Federal Energy Administration (FEA), a predecessor organization to EIA. PIES/MEFS was a static model, solving for one designated future year at a time. For its initial use in the Project Independence Blueprint, it provided forecasts for 1977, 1980, and 1985. For the 1981 Annual Report to Congress, the system projected 1985, 1990, and 1995.

The core model of PIES/MEFS was a single linear program of regional fuel supply, transportation, conversion, and end-use demand activities. This linear program optimized by solving for the configuration of fuel production and transportation to meet demand at the least cost to consumers. PIES/MEFS solved for a supply-demand equilibrium in energy markets by iterating between the linear program and a reduced-form representation of end-use demand models. From the linear program were derived the marginal, or shadow, prices for fuel delivered to the end-users by sector and region. The reduced-form demand representation was evaluated at these prices, the new end-use demands entered into the linear program, and the program reoptimized. This process continued until the end-use prices and demands were not changing between iterations, within a specified tolerance.

---

[1]This section draws from papers written at the committee's request. Susan Shaw, Energy Information Administration, wrote the material on PIES, MEFS, IFFS, LEAP and STIFS, and Eric Petersen, Office of Policy, Planning and Analysis, wrote the material on Fossil2.

As the model evolved over time, a number of special features were added to reflect regulatory policies or to ensure that certain end-use prices from the linear program were average or regulated prices, rather than strict marginal prices. Other features were added to ensure intertemporal consistency between the time periods of the model. These capabilities were incorporated in either the linear program directly or in the program that controlled the equilibrium and determined convergence.

As a modeling system, PIES/MEFS encompassed a host of satellite models--coal, oil, natural gas, synthetic fuels, refinery, electric utility, end-use demand, and macroeconomic. Each of these satellite models produced the necessary coefficients and objective function costs for the linear program and incorporated sector-specific features as required. Other models used the results of PIES/MEFS to perform macroeconomic and distributional analysis. This structure served to organize the data and allocate responsibilities for the modeling activities.

INTERMEDIATE FUTURE FORECASTING SYSTEM

The current EIA integrated modeling system is the Intermediate Future Forecasting System (IFFS), developed in 1982. IFFS partitions the energy system into fuel supply, conversion, and end-use demand sectors and then solves for a supply-demand equilibrium by successively and repeatedly invoking the modules that represent these sectors. The model solves annually, currently with a forecast horizon of 2010. The supply-demand equilibration is performed one year at a time, stepping forward to the next forecast year when the equilibrium for one year is complete. Fundamental assumptions for the modeling system are the assumptions for the world crude oil price and a baseline macroeconomic forecast.

The fuel supply modules of IFFS encompass all the activities necessary to produce, import, and transport the fuel to the end user, thus computing the domestic production and the regional end-use prices necessary to meet end-use demand. Each of the end-use demand modules compute the fuel requirements for the sector by region, based upon the regional end-use prices of all competing fuels, as well as other factors such as economic variables and technology characteristics. The electricity module, as a conversion module, simulates an input of fuel based on relative prices and technology characteristics as well as generates and prices electricity as an output.

Within IFFS, the primary interfaces between the modules are the regional end-use prices and demands for each fuel. Each fuel supply or end-use demand module is called in sequence and solves assuming all other variables in the energy markets are held constant. That is, the coal module solves for the production and end-use prices of coal, assuming a slate of demands for coal and assuming that all other sectors are fixed. Any module that uses the coal prices would then use these new prices to compute demand the next time the module is executed. As the model solves, the modules are called in sequence and percentage differences between iterations for all end-use, regional prices and demands are computed. When differences are within the user-specified tolerance, convergence is declared for that forecast year and the solution of the next forecast year begins.

Some attempt is made in IFFS to utilize convergence characteristics of particular modules. For example, the sensitivity of the natural gas price to the level of demand and the concomitant sensitivity of gas demand in certain sectors to the price is well recognized. So the electricity module computes a derived demand curve for natural gas, explicitly representing the demand for gas at a variety of prices, for both the electricity and gas modules to deal more effectively with convergence.

Due to the partitioning of the energy markets and the specific implementation of the modules, it is straightforward to execute any subset of the modules or a single module in IFFS or to substitute a module that meets a minimum interface requirement. In addition, the modular nature of IFFS readily allows each sector of the energy market to be represented with the methodology deemed most suitable to that sector, allowing for a more natural representation of each market. IFFS currently contains a mix of simulation, process, econometric, and optimization methods within the various sectors. It also allows each module to vary the depth and breadth of its coverage of the sector. As examples, there is a variation in whether the fuel supply modules explicitly include a representation of the fuel transportation and in what regional definition is used for fuel supply.

Like the PIES/MEFS system, IFFS utilizes a number of satellite models or analytical procedures in order to represent the energy market. Although IFFS encompasses more of the modeling directly, there still remain exogenous components. This is due either to the initial design of the exogenous models or to the necessity for increased flexibility of the component. Examples include a world coal trade model that provides export projections, an oil market model for world crude oil price forecasts, and a procedure for deriving natural gas import forecasts.

LONG-TERM ENERGY ANALYSIS PROGRAM

From 1979 to 1981, EIA utilized the Long-Term Energy Analysis Program (LEAP) for long-range forecasting to the year 2020. LEAP was EIA's configuration of the Generalized Equilibrium Modeling System (GEMS), originally developed by the Stanford Research Institute (currently, SRI International) and now with Decision Focus, Inc. This is utilized by a number of organizations, configured to suit their particular purposes.

LEAP/GEMS solves for a supply-demand equilibrium in a way fundamentally similar to IFFS with prices and quantities of the various types of energy being computed by modules that represent production, raw material transportation, conversion, final product transportation, and end-use energy consumption. Within the overall system, the order of solution is directional, prices flowing from supply to end-use demand and quantities flowing in reverse. Thus, it solves for an equilibrium by a recursive technique. One difference between LEAP/GEMS and IFFS is the method of solving through time. Each module solves for all forecast years at a time in LEAP/GEMS, attaining an equilibrium for all years simultaneously and enabling some modules to utilize perfect foresight over the forecast horizon.

LEAP, or any GEMS-derived model, segments the energy system by separating all supply, transportation, conversion, and end-use processes. Each of these activities is defined as a node, and a network describing the flows of all information between nodes must be explicitly drawn. Every regional activity, such as coal supply by region, would also be a separate node. At all decision points in the network, there are allocation nodes that represent the fuel or technology choice or the regional mix for different supply options. The allocation nodes use market share algorithms with market share coefficients, price premiums, behavioral lag coefficients, and initial year market shares.

One feature of the GEMS system is a library of generic models, from which one can choose in building a representation of the energy system. These generic models include a simple and a complex conversion process, an allocation process, a primary resource process, a end-use demand process, and a transportation process. In building a model using GEMS, a user draws the network by selecting a generic model for each node, defining all the input and output links to other nodes, and specifying all necessary data to characterize the node specifically. It is the data that distinguishes, for example, a residential electric heat pump node from an industrial machine drive node. If a model builder wishes to add new generic models to the system, it can be done. As an example, for LEAP, EIA created a separate coal supply module as distinguished from the oil and gas supply modules, believing that a single primary resource process could not adequately represent such dissimilar fuel supplies.

SHORT-TERM INTEGRATED FORECASTING SYSTEM

A very different modeling system has been used by EIA since 1979 for short-term forecasting and related analysis. The short-term energy outlook of EIA presents a two-year, quarterly forecast of energy supply and demand, produced using the Short-Term Integrated Forecasting System (STIFS). Primary inputs to STIFS include assumed prices for crude oil and natural gas, macroeconomic indicators, assumed weather patterns, and current data on the energy system, including inventory levels.

STIFS is a national representation of energy markets. Consumption for each of the major fuels is computed based upon relative end-use prices and recent trends. Domestic crude oil production is a function of the assumed crude oil price. The consumption forecasts and trends for fuel production, imports, and stock levels are integrated into an energy balance for each of the fuels. Thus STIFS does not account for all feedbacks of energy prices or consumption on production. Being a short-term system, STIFS also does not account for capital stock changes.

FOSSIL2

The integrating analysis tool for the National Energy Strategy is a large-scale model of the U.S. energy system called Fossil2. Fossil2 is a dynamic simulation model of U.S. energy supply and demand designed to project the long-term (30 to 40 year) behavior of the U.S. energy system. The model structure, which includes all energy producing and consuming sectors, simulates the marketplace in a series of dynamic stocks and flows; the stocks include energy production facilities (e.g., oil fields), energy transformation facilities (power plants) and energy consuming entities (e.g., houses, vehicles), while the flows include energy, prices and information.

The Fossil2 model can be characterized as an equilibrium energy market model, as energy markets "clear" over time through feedback among such factors as prices, demand, conservation investments, production costs and production capacity. The model clears markets for each iteration period, which is a variable within the model that is typically set to one quarter year. The model uses System Dynamics, a methodology that represents system behavior through differential and integral equations.

In Fossil2 the demand for energy is determined in a "least cost/energy services" framework of total U.S. energy demand. Following this approach, the model first projects the demand for energy services (heat, light, steam, shaft power) in each end-use category, and then calculates the share of service demand captured by end-use technologies.

For most energy service categories, there are several fuels that can provide the required energy services--in addition to "conservation." In a few categories (such as lighting or appliances), only electric energy services can be used. For those where there are choices of fuels, a least-cost algorithm based on consumer choice theory is used in Fossil2 to determine the market share for different fuel- using categories of new energy equipment.

The energy supply sectors of the Fossil2 model represent the decisions that lead to the commitment to new production capacity, the operation of existing production capacity, and the setting of energy prices for oil, gas, coal and electricity. Energy producers choose to invest in production technologies that maximize the industry's rate of return (or minimize the average cost of production), subject to environmental constraints (for example, SO_2 restrictions). The sectors keep account of production capacity and assets, and calculate energy prices in accordance with the rules estimated to be followed in each industry. These rates then feed into the demand sector, helping to determine current and future growth in energy demand.

APPENDIX F

MEETINGS AND ACTIVITIES

July 31 & August 1, 1990

"The National Energy Strategy and the National Energy Modeling System"
Henson Moore, Deputy Secretary of Energy, and Linda G. Stuntz, Deputy Under Secretary, Policy, Planning and Analysis, U.S. Department of Energy (DOE).

"The Development and Operation of the NEMS: An EIA Perspective"
Calvin A. Kent, Administrator Designate, Energy Information Administration (EIA).

"Reference Case for the National Energy Strategy"
Eric Petersen, DOE Office of Policy, Planning and Analysis.

"Current Configuration and Applications of the NEMS"
W. Calvin Kilgore, Director, EIA Office of Energy Markets and End Use.

"Use of Energy Models and Data Systems at EIA"
Lawrence A. Pettis, Deputy Administrator, EIA.

"Requirements Analysis for the NEMS"
C. William Skinner, Technical Assistant to the Administrator, EIA.

September 20 & 21, 1990

"Update on NES and Key Policy Issues"
Linda G. Stuntz, Deputy Under Secretary, Policy, Planning and Analysis, DOE.

"Context for the Analysis of Key Policy Issues"
   --Descriptions of Options
   --Definition of the "Reference Case"
   --Conceptual basis for integrated analysis
   --Choice of issues to "demonstrate" characteristics
      & capabilities of models applied to the current NES effort
Eric Petersen, DOE Office of Policy, Planning and Analysis.

"Sectoral Energy Demand: PC-AEO Models"
John D. Pearson, Director, Energy Analysis & Forecasting Division, EIA Office of Energy Markets & End Use.

"Energy Supply - Coal & Electricity: NCM & ARGUS"
Mary J. Hutzler, Director, Electric Power Division, EIA Office of Coal, Nuclear, Electric & Alternate Fuels.

"Energy Supply - Oil & Gas" GAMS & PROLOG"
Susan Shaw, Analysis & Forecasting Branch, Reserves & Natural Gas Division, EIA Office of Oil and Gas.

"Integration Model: FOSSIL 2"
Roger Nail, AES Corporation, Arlington, VA.

"Oil Market Simulation Model"
Erik Kreil, International/Contingency Information Division,
EIA Office of Energy Markets & End Use.

"DRI Macroeconomic Model"
Ronald Earley, Economics & Statistics Division, EIA Office of Energy Markets and End Use.

"Wrap-up on Current NES Analysis Effort"
Robert C. Marlay, Acting Director, DOE Office of Program Review & Analysis.

"Update on NEMS Development/Look-Ahead"
Calvin A. Kent, Administrator, EIA.

November 6, 1990

Preparatory Meeting, preceding Secretary's meeting with the Chairman and Committee Members on November 9, 1990 to exchange impressions regarding the application of existing models and data by DOE & EIA to the ongoing national energy strategy exercise.

November 8 and 9, 1990

Committee meeting on November 8-9, and Meeting with Secretary Watkins by the delegation from the committee on November 9, 1991, from 10 a.m. to 12 noon.

January 17 and 18, 1991

"Modeling Energy, Economic and Environmental Interactions: Applications to CO_2 Emissions Abatement"
Rich Richels, Program Manager, Environmental Risk Analysis, Electric Power Research Institute (EPRI)

"Issues and Considerations in Developing Long-Range Modeling Capabilities at DOE"
Dale M. Nesbitt, Vice President, Decision Focus Inc.

"Energy Modeling, Forecasting and Planning in the Pacific Northwest"
Jim Litchfield, Northwest Power Planning Council and Sue Hickey, Assistant Administrator for Energy Resources, Bonneville Power Administration.

"Energy Modeling at the California Energy Commission"
Daniel Nix, Deputy Director for Energy Forecasting and Planning, California Energy Commission (CEC).

"Development and Application of the ELFIN Model"
Dan Kirshner, Environmental Defense Fund.

"Energy Demand Forecasting Models Developed by EPRI"
Philip Hummell, Customer Systems Division, EPRI.

"Long-term Policy Issues on Energy"
Robert C. Marlay, Acting Director, Office of Program Review and Analysis, DOE.

February 28 and March 1, 1991

"NEMS Requirements for Responsive Policy Analysis"
Robert C. Marlay, Office of Policy, Planning and Analysis, DOE.

"Examples of Analysis and Decision Processes at the Power Planning Council"
James W. Litchfield, Northwest Power Planning Council.

"Addressing Value Issues in Modeling Decisions"
Ralph L. Keeney, Professor, University of Southern California.

April 18 and 19, 1991

"Status of the NEMS Project at EIA"
John Holte, NEMS Project Office, EIA.

"Modularity as it relates to NEMS"
Susan Shaw, NEMS Project Office, EIA.

"Archiving of Models"
Douglas R. Hale, Director, Quality Assurance Division, EIA.

June 6 and 7, 1991

"Modeling Activities and the NEMS Tie-in in the Office of Conservation and Renewable Energy"
Henry Kelly, Office of Conservation & Renewable Energy, DOE.

"Long-Range Modeling"
Jae Edmonds, Battelle Pacific NW Labs, Washington, D.C.

"Long-Range Forecasting"
Lester Lave, Carnegie-Mellon University.

"100 Year Period--A Century of Uncertainty"
David Gray, David Morrison and Glen Tomlinson, Mitre Corporation, McLean, Virginia.

July 15 and 17, 1991

Writing Group of the committee worked on draft of report.

July 25 and 26, 1991

"EIA Data and Analytical Tools: End-Use and Efficiency"
Eric Hirst, Oak Ridge National Laboratories, Oak Ridge, Tennessee.

"What Makes the Office of Technology Assessement Work?"
John H. Gibbons and Peter Blair, Office of Technology Assessment, Washington, D. C.

October 4, 1991

Exchange of views with Secretary of Energy
Staff and committee members.

REFERENCES AND BIBLIOGRAPHY

Alliance to Save Energy. 1983. Industrial Investment in Energy Efficiency: Opportunities, Management Practices and Tax Incentives. Washington, D.C.

Ayres, R. U. 1978. Resources, Environment, and Economics: Applications of the Materials/Energy Balance Principle. Wiley-Interscience, New York.

The AES Corporation. 1990. An Overview of the FOSSIL2 Model: A Dynamic Long-term Policy Simulation Model of U.S. Energy Supply and Demand, prepared for the U.S. Department of Energy, Office of Policy and Evaluation (July 6). DE-ACO1-89PE79041. The AES Corporation: Arlington, Virginia.

Ballard, C.L. and L.H. Goulder. 1985. Consumption Taxes, Foresight, and Welfare: A Computable General Equilibrium Analysis. In J. Piggott and J. Walley (eds.), New Developments in Applied General Equilibrium Analysis, pp. 253-82. Cambridge: Cambridge University Press.

Berndt, B.R. 1991. The Practice of Econometrics: Classic and Contemporary. Reading, Massachusetts: Addison-Wesley Publishing Co.

Bohi, D.R. and M.A. Toman. 1984. Analyzing Non-renewable Resource Supply. Washington, D.C.: Resources for the Future.

Bohi, D.R.B. 1981. Analyzing Demand Behavior: A Study of Energy Elasticities. Baltimore: Johns Hopkins University Press.

Bookout, J.F. 1989. Two Centuries of Fossil Fuel Energy. Remarks to the International Geological Congress, July 10, Washington, D.C.

Borges, A.M. and L.H. Goulder. 1984. Decomposing the Impact of Higher Energy Prices on Long-term Growth. In H.E. Scarf and J.B. Shoven (eds.), Applied General Equilibrium Analysis, pp. 319-62. Cambridge: Cambridge University Press.

Boyd, G., J.F. McDonald, M. Ross, and D.A. Hanson. 1986. Effects of the Changing Composition of U.S. Manufacturing Production on Energy Demand. Report No. CONF-8603105. Palo Alto, California: Energy Modeling Forum.

Brock, H.W. and D.M. Nesbitt. May 1977. Large Scale Energy Planning Models: A Methodological Analysis, Chapter 1. Prepared for The National Science Foundation, Office of Policy Research and Analysis, Washington, D.C.

Canadian Energy Research Institute (CERI) and Decision Focus Incorporated. 1990. A Proposal for a World Gas Trade Program Based on a World Gas Supply, Transportation and Demand Model and Data Base. (April).

Chao, H.P., B.R. Judd, P.A. Morris, and S.C. Peck. 1985. Analysing Complex Decisions for Electric Utilities, Electric Power Research Institute (EPRI), Palo Alto, California.

Chern, W.S., C.A. Gallagher, R.C. Tepel and J.L. Trimble. 1982. An Integrated System for Forecasting Electric Energy and Load for States and Utility Service Areas, Oak Ridge National Laboratory Report TM-7947. Oak Ridge, Tennessee.

Cohen, J.E. 1986. Population Forecasts and Confidence Intervals for Sweden: A Comparison of Model-Based and Empirical Approaches. Demography 23: 105-126.

Conti, J., and S. Shaw. 1988. Modeling and Forecasting Energy Markets with the Intermediate Future Forecasting System, Operations Research (May-June).

Conti, J. 1991. Regionality in the National Energy Modeling System. NEMS Project Office, Issue Paper, Energy Information Administration (May 21), Washington, D.C.

Cowing, T. and D. McFadden. 1984. Microeconomic Modeling and Policy Analysis. Academic Press.

Decision Focus Incorporated (DFI). The California Petroleum Economy Model CAPE Model Structure and Methodology. Los Altos, California.

Denning, P.J. 1991. Modeling Reality. American Scientist, Volume 78:495-498.

Difiglio, C.D., K.G. Duleep, and D.L. Greene. 1990. Cost Effectiveness of Fuel Economy Improvements. Energy Journal, Vol. 11, No. 1, pp. 65-86.

Diwekar, U.M. and E.S. Rubin. 1991. Stochastic Modeling of Chemical Processes. Computers and Chemical Engineering, Vol. 15, No. 2, pp. 105-114.

Doblin, C.P. 1984. Changing Structure of Industry and its Impact on Energy Requirements. In Weyant, J.P. and D.B. Sheffield (eds.). Energy Industries in Transition 1985-2000, Part 2, pp. 1325-1340. Washington, D.C.: International Association of Energy Economists.

Dubin, J. 1985. Consumer Durable Choice and the Demand for Electricity. New York: North-Holland Publishing Co.

Dubin, J. and D. McFadden. 1984. An Econometric analysis of Residential Electric Appliance Holdings and Consumption. Econometrica, Vol. 52, No. 2, pp. 345-362.

The Economist. 1991. Schools Brief: The next 100 years (March 9), pp 74-75.

Edmonds, J.A. 1991. Overview and Assessment of the State of Economic/Energy/Environment Policy Modeling and Modeling Research Agendas. Panel remarks in D.O. Wood and Y. Kaya, Proceedings of the Workshop on Economic/Energy/Environmental Modeling for Climate Policy Analysis. Center for Energy Policy Research and Sloan School of Management, Massachusetts Institute of Technology, Cambridge, Massachusetts.

Edmonds, J. and D.W. Barns. 1990. Estimating the Marginal Cost of Reducing Global Fossil Fuel CO_2 Emissions. Report PNL-SA-18361. Pacific Northwest Laboratory, Washington, D.C.

Egan, J. 1990. Is Energy Strategy Falling Victim To A War Between Rival Computer Models? The Energy Daily (August 10).

Electric Power Research Institute (EPRI). Residential Sector Technology Characterization, Palo Alto, CA.

Electric Power Research Institute (EPRI). 1977. Report EA 0837. Initiation of Integration. EPRI Reserch Reports Center, Palo Alto, CA.

Energy Information Administration (EIA). 1991. Letter from Calvin Kent, Administrator, to Peter Johnson, chairman of the committee.

Energy Information Administration (EIA). 1990a. Requirements Analysis For a National Energy Modeling System, paper prepared by an Energy Information Administration Working Group. Washington, D.C.: U.S. Department of Energy. (July 2).

Energy Information Administration (EIA). 1990b. Directory of Energy Information Administration Models 1990. DOE/EIA-0293(90). Washington, D.C.: U.S. Department of Energy.

Energy Information Administration (EIA). 1990c. Annual Energy Outlook with Long-Term Projections. DOE/EIA-0383(90). Washington, D.C.: U.S. Government Printing Office.

Energy Information Administration (EIA). 1990d. Annual Energy Outlook with Quarterly Projections. DOE/EIA-0202(90/2Q). Washington, D.C.: U.S. Government Printing Office.

Energy Information Administration (EIA). 1990e. National Energy Modeling System (NEMS): A Comparison of Requirements with Current Capabilities and Issues in the Design of a New System, Energy Information Administration Working Group. Washington, D.C.: U.S. Department of Energy. (September 19).

Energy Information Administration (EIA). 1990f. Improving Technology: Modeling Energy Futures for the National Energy Strategy. SR/EIA/90-7, Washington, D.C. U.S. Department of Energy. (October).

Energy Information Administration (EIA). 1990g. PC-AEO Forecasting Model for the Annual Energy Outlook 1990, Model Documentation (March). Washington, D.C.

Energy Information Administration (EIA). 1989. Directory of Energy Data Collection Forms. Forms in Use as of October 1989. DOE/EIA-0249(89). Washington, D.C.: U.S. Department of Energy.

Energy and Environmental Analysis, Inc. (EEA). 1982. Industrial Sector Technology Use Model, Vols. 1-8, Arlington, Virginia.

Fisher, F. M. and C. Kaysen. 1962. A Study in Econometrics: The Demand for Electricity in the United States, North Holland Publishing Co.

Freedman, D. 1981. Are Energy Models Credible? pp. 93-116 in Gass, S. I. (ed.) Validation and Assessment of Energy Models. National Bureau of Standards Special Publication 616.

Frey, H.C. and E.S. Rubin. 1991. Stochastic Modeling of Coal Gasification Combined Cycle Systems. Task 3 Topical Report. Contract No. DE-AC21-88MC24248. Morgantown, West Virginia: U.S. Department of Energy (May).

Gas Research Institute (GRI). undated. Contracting and Licensing: Policy and Practice. Contract Administration Department, Chicago, Illinois.

Goett, A.A. and D. McFadden. 1984. The Residential End-Use Energy Planning System: Simulation Model Structure and Empirical Analysis. In: J.R. Moroney, Ed., Advances in the Economics of Energy and Resources, Vol. 5, Greenwich, Connecticut: JAI Press: 153-210.

Goulder, L. Lawrence Berkeley Laboratory. Residential Sector Technology Characterization.

Gouse, W.S. 1991. Long-Term Energy Conversion Research and Development Needs. McLean, Virginia: The Mitre Corporation.

Granger M., M. Henrion, and M. Small. 1990. Uncertainty: A Guide to Dealing with Uncertainty in Quantitative Risk and Policy Analysis. New York: Cambridge University Press.

Greenberger, M., et. al. 1976. Models in the Policy Process: Public Decision Making in the Computer Era. Chapter 1. Russell Sage Foundation, New York.

Griffin, J.M. 1982. Pseudodata: A Synthesis of Econometric and Process Modeling Methodologies. In: R. Amit and M. Auriel (Eds.), Perspectives on Resource Policy Modeling: Energy and Minerals, Ballinger Publishing Co., Cambridge, Massachusetts.

Griffin, J.M. 1978. Joint Production Technology: the Case of Petrochemicals. Econometrica (March):379-396.

Griffin, J.M. 1977. Long-run Production Modeling with Pseudodata: Electric Power Generation. Bell J. (Spring):112-127.

Griffin, J.M. 1985. OPEC Behavior: A Test of Alternative Hypothesis. American Economic Review, Vol. 75, No. 5.

Halvorson, R. 1978. Econometric Models of U.S. Energy Demand. Lexington: D.C. Heath Publishing Co.

Hausman, J.A. 1981. Validation and Assessment Procedures for Energy Models. In Gass, S.I. (ed.) Validation and Assessment of Energy Models, pp. 79-86 . National Bureau of Standards Special Publication 616.

Hausman, J.A. 1979. Individual Discount Rates and the Purchase and Utilization of Energy-Using Durables, Bell J. of Economics 10(1):33.

Heyde, C.C. and J.E. Cohen. 1985. Confidence Intervals for Demographic Projections Based on Products of Random Matrices. Th. Pop. Biology 27:120-153.

Hickman, B.G., H.G. Huntington, and J.L. Sweeney. 1987. Macroeconomic Impacts of Energy Shocks. Amsterdam: North-Holland Publishing Co.

Hickman, B.G., H.G. Huntington, H.G., and J.L. Sweeney (eds.). 1987. Macroeconomic Impacts of Energy Shocks. Contributions to Economic Analysis Series, No. 163. Amsterdam, Oxford and Tokyo: North-Holland Publishing Co.

Hirst, E., and M. Schweitzer. 1990. Electric-Utility Resource Planning and Decision-Making: The Importance of Uncertainty. Risk Analysis, Vol. 10. No. 1.

Hirst, E. 1990. Balancing the Scales: Toward Parity in Electric Supply and Demand Data.. Oak Ridge National Laboratory, The Electric Journal: 28-33.

Hirst, E. 1989. Comparison of EIA Data Collections: Electricity Supply and Demand, (October). Energy Division, Oak Ridge National Laboratory, Oak Ridge, Tennessee.

Hodges, J. 1987. Uncertainty, policy analysis and statistics. Statist. Science. 2:259-273.

Hogan, W.W. and D.W. Jorgenson. 1991. Productivity Trends and the Cost of Reducing CO_2 Emissions. Energy Journal, Vol. 12, No. 1, pp. 67-85.

Hogan, W.W. and J.P. Weyant. 1983. Methods and Algorithms for Energy Model Composition: Optimization in a Network of Process Models. In: Benjamin, L. (ed.), Energy Models and Studies. Studies in Management Science and Systems Series, Vol. 9. New York, Amsterdam and Oxford: North-Holland Publishing Co.

Hogan, W.W. 1989. A Dynamic Putty-Semi-putty Model of Aggregate Energy Demand. Energy Economics, Vol.11, No. 1, pp. 53-69.

Holdren, J.P. 1990. Energy in Transition. Scientific American (September), pp. 157-163.

Holte, J. 1991. Near-Term, MidTerm and Long-Term Forecasting in the National Energy Modeling System, NEMS Project Office, Issue Paper, Energy Information Administration, (May 21), Washington, D.C.

House, P.W., and P. McLeod, P. 1977. Large Scale Models for Policy Evaluation. Wiley-Interscience, New York.

Houthakker, H.S. and L.D. Taylor. 1970. Consumer Demand in the United States, Second Edition, Cambridge: Harvard University Press.

Houthakker, H.S., P.K. Verleger, and D.P. Sheehan. 1974. Dynamic Demand Analysis for Gasoline and Residential Electricity. American Journal of Agricultural Economics, 56:2.

Hubbard, H. 1991. The Real Cost of Energy. Scientific American (April). Vol. 264, No. 4:36-42.

Huntington, H.G. and G.E. Schuler, Jr. 1990. North American Natural Gas Markets: Summary of an Energy Modeling Forum Study. Energy Journal, Vol. 11, No. 2, pp. 1-21.

Jaske, M.R. 1990. Air Pollution Projection Methodologies: Integrating Emission Projections with Energy Forecasts. California Energy Commission. Asilomar Conference Center, Pacific Grove, California.

Jorgensen, D. and P. Wilcoxen. 1990a. Environmental Regulation and U.S. Economic Growth. The Rand Journal of Economics 21 (2):314-340.

Jorgenson, D.W. and P.J. Wilcoxen. 1990b. Intertemporal General Equilibrium Modeling of U.S. Environmental Regulation, Journal Of Policy Modeling, Vol. 12, No. 4:715-744.

Keeney, R.L. 1988. Structuring Objectives for Problems of Public Interest. Operations Research, Vol. 36, No. 3, (May-June) 396-405.

Keeney, R.L. 1987. An Analysis of the Portfolio of Sites to Characterize for Selecting a Nuclear Repository, Risk Analysis, Vol. 7, No. 2: 195-218.

Keeney, R.L. 1988. Building Models of Values. Invited Review, European Journal of Operational Research, Vol. 37: 149-157

Keeney, R.L. 1987. A Multiattribute Utility Analysis of Alternative Sites for the Disposal of Nuclear Waste, Risk Analysis, Vol. 7, No. 2: 173-194.

Kneese, A.V. and J.L. Sweeney. 1991. Handbook of Natural Resources and Energy Economics, Vol. 3, North-Holland Publishing Co., Forthcoming. In particular for nonrenewable resource models see: Newberry, D. and L. Karp, Intertemporal Consistency Issues in Depletable Resources; Slade, M. Kolstad, C. and R. Winer, Buying Energy and Nonfuel Minerals: Final, Derived, and Speculative Demand; Epple, D. and J. Londregan, Strategies for Modeling Exhaustible Resource Supply.

Koreisha, S., and R. Stobaugh, An appendix ("Limits to Models") in Energy Future, Eds. R. Stobaugh and D. Yergin. 1979. pp.234-339.

Leamer, E.E. 1983. Let's Take the Con Out of Econometrics. Amer. Econ. Review. 73:31-43.

Lucas, R.E. and T.J. Sargent (eds.).1981. Rational Expectations and Econometric Practice. Minneapolis: University of Minnesota Press.

MacAvoy, P. 1982. Crude Oil Prices as Determined by OPEC and Market Fundamentals. Cambridge: Ballinger Publishing Co.

Malliaris, A.G. and Brock, W.A. (1982) Stochastic Methods in Economics and Finance. New York: North-Holland.

Manne, A. and Richels, R.G. 1990. Buying greenhouse insurance. Preprint.

Manne, A.S., Richels, R.G. and W.W. Hogan. 1990. CO2 Emission Limits: An Economic Cost Analysis for the USA. Energy Journal, Vol. 11, No. 35, p. 51.

Manne, A.S., So, K.C., Weyant, J.P., Kydes, A.S. and D.M. Geraghty. 1985. OTM: An International Oil Trade Model. In Energy Markets in the Longer-term: Planning under Uncertainty, pp. 241-44. New York: Elsevier Publishing Co.

Marlay, R. 1991. Presentation to the NEMS Committee at its February, 1991 Meeting (see Appendix F).

McFadden, D. 1983. Econometric Models of Probabilistic Choice. In C. Manski and D. McFadden (eds.). Structural Analysis of Discrete Data. Cambridge: MIT Press.

McMahon, J.E. 1986. The LBL Residential Energy Model. Applied Science Division. Lawrence Berkeley Laboratory, Berkeley, California.

Merrow, E.W., Phillips, K.E. and Myers, C.W. 1981. Understanding Cost Growth and Performance Shortfalls in Pioneer Process Plants. Report R-2569- DOE. Rand Corporation, Santa Monica, Calif.

Morgan, M.G., and M. Henrion. 1990. Uncertainty: A Guide to Dealing with Uncertainty in Quantitative Risk and Policy Analysis, Cambridge University Press.

Muth, J.F. 1981a. Estimation of Economic Relationships Containing Latent Expectations Variables. In R.E. Lucas and T.J. Sargent (eds.), Rational Expectations and Econometric Practice, pp. 321-28. Minneapolis: University of Minnesota Press.

Muth, J.F. 1981b. Rational Expectations and the Theory of Price Movements. In R.E. Lucas and T.J. Sargent (eds.). Rational Expectations and Econometric Practice, pp. 3-22. Minneapolis: University of Minnesota Press.

National Academy of Sciences (NAS). 1991. Policy Implications of Greenhouse Warming. Washington, D.C.: National Academy Press.

National Research Council (NRC). 1991a. Development of the National Energy Modeling System. First Advisory Report. Committee on the National Energy Modeling System. Washington, D.C.: Energy Engineering Board.

National Research Council (NRC). 1991b. An Evaluation of the Department of Interior's 1989 Assessment Procedures. Washington, D.C.: National Academy Press.

National Research Council (NRC). 1991. Improving Information for Social Policy Decisions: The Uses of Microsimulation Modeling, Vol. 1, Review and Recommendations. Washington, D.C.: National Academy Press.

National Research Council (NRC). 1990. Fuels to Drive Our Future. Washington, D.C.: National Academy Press.

National Research Council (NRC). 1977. Energy Consumption Measurement. Data Needs for Public Policy. Committee on Measurement of Energy Consumption. Assembly of Behavioral and Social Sciences. Washington, D.C.

Nelson, C.R. and S.C. Peck. 1985. The NERC Fan: A Retrospective Analysis of NERC Summary Forecasts. Journal of Business and Economic Statistics 3:3.

Nelson, C.R., S.C. Peck, and R.G. Uhler. The NERC Fan in Retrospect and Lessons for the Future. Energy Journal, Volume 10, No. 2, pp. 91-107.

Nesbitt, D.M. 1983. The Economic Foundation of Generalized Equilibrium Modeling. Los Altos, California: Decision Focus Incorporated.

Northwest Power Planning Council (NPCC). 1991. Northwest Conservation and Electric Power Plan, Volumes 1 and 2. Portland, Oregon.

Office of Technology Assessment (OTA). 1989. Statistical Needs for A Changing U.S. Economy. Background Paper, OTA-BP-E-58. U.S. Congress. Washington, D.C.: U.S. Government Printing Office.

Office of Technology Assessment (OTA). 1985. New Electric Power Technologies: Problems and Prospects for the 1990s. U.S. Congress: Washington, D.C. U.S Government Printing Office.

Office of Technology Assessment (OTA). 1989. Background Paper OTA-BP-E-57. Energy Use and the U.S. Economy. U.S. Congress. Washington, D.C.: U.S. Government Printing Office.

Ottinger, R.L., et.al. 1990. Environmental Costs of Electricity. Prepared for New York State Energy Research and Development Authority and U.S. Department of Energy, pp. 477-557.

Pacific Gas and Electric Company and Natural Resources Defense Council. July 16, 1990. Joint Comments of The Natural Resources Defense Council and the Pacific Gas and Electric Company on the U. S. Department of Energy's National Energy Strategy. San Francisco, California.

Peck, S.C., T.J. Teisberg, "CETA: A Model for Carbon Emissions Trajectory Assessment", Electric Power Research Institute, Palo Alto, California.

Peck S.C., D.K. Bosch, and J.P. Weyant. 1988. Industrial Energy Demand: A Simple Structural Approach. Resources and Energy 10.

Powell, S.G. 1990. The Target Capacity-Utilization Model of OPEC and the Dynamics of the World Oil Market. The Energy Journal. Energy Economics Educational Foundation, Inc., Vol. II, No. 1.

Reister, D.B. 1984. Simple Model of the Greenhouse Effect. In J. P. Weyant and D. B. Sheffield (eds.), Energy Industries in Transition 1985-2000, Part 1, pp. 563-577. Washington, D.C.: International Association of Energy Economists.

Ross, M. 1984. Energy-conservation Investment Practices of Large Manufacturers. In J. P. Weyant and D. B. Sheffield (eds.). Energy Industries in Transition 1985-2000, Part 2, pp. 977-990. Washington, D.C.: International Association of Energy Economists.

Rubin, D. 1990. A New Perspective. In K. W. Wachter and M. L. Straf (eds.), The Future of Meta-analysis, pp. 155-166. Russell Sage Foundation, New York.

Rubin, E.S. and U.M. Diwekar. 1989. Stochastic Modeling of Coal Gasification Combined Cycle Systems. Task 1 Topical Report. Contract No. DE-AC21-88MC24248. Morgantown, West Virginia: U.S. Department of Energy.

Ruderman, H., M.D. Levine, and J.E. McMahon. 1987. The Behavior of the Market for Energy Efficiency in Residential Appliances Including Heating and Cooling Equipment, The Energy Journal, Vol. 8, No. 1.

Schwartz, P. 1990. Accepting the Risk in Forecasting. The New York Times FORUM.

Shaw, S.H. undated. Modeling Energy Markets with the Intermediate Future Forecasting System. Forthcoming in a Handbook of Engineering-Economic Modeling.

Sims, C.A. 1988. Uncertainty Across Models, Amer. Econ. Review 78: 163-167.

Sims, C.A. 1982. Foundations of Modeling. In M. Hazewinkel, and A. H. G. R. Kan Current Developments in the Interface: Economics, Econometrics, Mathematics. D. Reidel, Boston.

Sims, C.A. 1984. Specification, Estimation, and Analysis of Macroeconometric Models. J. Money, Credit and Banking (18):121-126.

Skinner, W.C. 1991. A View of the Future National Energy Modeling System," NEMS Project Office, Issue Paper, Energy Information Administration (May 21), Washington, D.C.

Spencer, G. 1989. Projections of the Population by Age, Sex and Race. U.S. Bureau of the Census: 1988 to 2080. Current Population Reports, Series P-25 #1018. U.S. Government Printing Office. Washington, D.C.

Squitieri, R. 1984. Industrial Electricity Demand 1984-2000. In J. P. Weyant and D. B. Sheffield (eds.). Energy Industries in Transition 1985-2000, Part 2, pp. 1087-1101. Washington, D.C.: International Association of Energy Economists.

Star, C. and M. Searl. 1990. Global Energy and Electricity Futures: Demand and Supply Alternatives. Energy Systems and Policy, Vol. 14:53-83.

Starobin, P. 1990. Foggy Forecasts. National Journal 22 (May), pp. 1212-1215.

Stobaugh R. and D. Yergin. Energy Future. New York: Random House.

Sweeney, J.L. and J.P. Weyant. 1979. The Energy Modeling Forum: Past, Present and Future. In P.N. Nemetz (ed.), Energy Policy: The Global Challenge. Institute for Research on Public Policy.

Sweeney, J.L. 1981. World Oil Preliminary Results from the Energy Modeling Forum Study. Report No. CONF-8105234, Energy Modeling Forum, Palo Alto, California.

Taylor, L.D., G.R. Blattenberger and R.K. Rennhack. 1984. Residential Energy Demand in the United States: Introduction and Overview of Alternative Models. In J. R. Moroney (ed.), Advances in the Economics of Energy and Resources, 5, JAI Press.

Toman, M.A. 1990. What Do We Know About Energy Security?. RESOURCES (Fall), No. 101, Resources For the Future, Washington, D. C.

Toman, M. 1991. The Economics of Energy Security: Theory, Evidence, Policy. In A.V. Kneese and J.L. Sweeney (ed.s). Handbook of Natural Resource and Energy Economics, Vol. 3, North-Holland Publishing Co. Forthcoming.

Trefil, J. 1990. Modeling Earth's Future Climate Requires Both Science and Guesswork (December), Vol. 21, No. 9. Smithsonian Magazine, Washington, D.C.

U.S. General Accounting Office (GAO). 1988. USDA's Commodity Program. The Accuracy of Budget Forecasts (April). Report to the Chairman, Subcommittee on Government Information, Justice, and Agriculture, Committee on Government Operations, House of Representatives. GAO/PEMD-88-8. Washington, D.C.

U.S. Department of Energy (DOE). 1991a. National Energy Strategy. Report DOE/S-0082P.

U.S. Department of Energy (DOE). 1991b. Fiscal Year 1992 Congressional Budget Request, Energy Information Administration, Vol.4. Washington, D.C. (January).

U. S. Department of Energy (DOE). 1990a. Interim Report. National Energy Strategy. A Compilation of Public Comments. DOE/S-0066P. Washington, D.C. (April).

U.S. Department of Energy (DOE). 1990b. Fiscal Year 1991 Congressional Budget Request, Fossil Energy Research and Development, Vol.4. Washington, D.C. (January).

U.S. Department of Energy (DOE). July 26, 1989. Statement of Admiral James D. Watkins, Secretary of Energy for Committee on Energy and Natural Resources, United States Senate, Washington, D.C.

U.S. Department of Energy (DOE). 1988. Assessment of Costs and Benefits of Flexible and Alternative Fuel Use in the U.S. Transportation Sector. DOE/PE-0080, Washington, D.C. (January).

Weyant, J.P. 1985. General Economic Equilibrium as a Unifying Concept in Energy-Economic Modeling. Management Science, Vol. 31, No. 5:548-563 (May).

Zimmerman, M.B. April 10, 1990. Assessing the Costs of Climate Change Policies. The Uses and Limits of Models. The Alliance to Save Energy. Washington, D.C.